The theory of nuclear magnetic relaxation
in liquids

THE THEORY OF

---

nuclear magnetic relaxation in liquids

JAMES McCONNELL

Senior Professor of Theoretical Physics, Dublin Institute for Advanced Studies

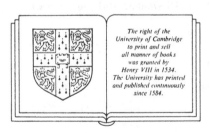

The right of the
University of Cambridge
to print and sell
all manner of books
was granted by
Henry VIII in 1534.
The University has printed
and published continuously
since 1584.

CAMBRIDGE UNIVERSITY PRESS

Cambridge

New York   New Rochelle

Melbourne   Sydney

CAMBRIDGE UNIVERSITY PRESS
Cambridge, New York, Melbourne, Madrid, Cape Town, Singapore, São Paulo, Delhi

Cambridge University Press
The Edinburgh Building, Cambridge CB2 8RU, UK

Published in the United States of America by Cambridge University Press, New York

www.cambridge.org
Information on this title: www.cambridge.org/9780521107716

First published 1987
This digitally printed version 2009

A catalogue record for this publication is available from the British Library

Library of Congress Cataloguing in Publication data
McConnell, J. (James)
Theory of nuclear magnetic relaxation in liquids.
Bibliography
Includes indexes.
1. Liquids–Magnetic properties.   2. Relaxation
phenomena.   3. Nuclear magnetic resonance.   4. Random
walks (Mathematics)   5. Diffusion processes.   I. Title.
II. Title: Nuclear magnetic relaxation in liquids.
QC145.4.M27M37   1987   530.4′2   86-26853

ISBN 978-0-521-32112-9 hardback
ISBN 978-0-521-10771-6 paperback

Contents

Contents

Preface

Shortly after the detection of nuclear magnetic resonance signals phenomenological theories of nuclear magnetic relaxation were proposed by Bloch (1946) and by Bloembergen, Purcell & Pound (1948). Their results constituted a framework for further investigations and provided reference points for comparison with later theories. Subsequent developments were the random walk theory of Torrey (1953) and the rotational diffusion theory of Abragam (1961). A further impetus towards providing a theory of nuclear magnetic relaxation was given by investigations on the rotational Brownian motion of an asymmetric molecule (Ford, Lewis & McConnell, 1979).

In spite of the rapidly increasing number of publications on nuclear magnetic resonance it is still difficult to find an intelligible and complete account of relaxation theory. It is the purpose of the present book to provide, as far as space allows, a self-contained account of this theory. Basic formulae are derived for relaxation times resulting from dipolar, quadrupolar, scalar and spin-rotational interactions, and from anisotropic chemical shift. Analytical expressions are then deduced for the relaxation times in the case of different molecular shapes. The level of knowledge presupposed in the reader would be found in undergraduate courses in mathematics, physics and chemistry. In the earlier chapters the notions of probability distributions and of random motion are introduced, and the method of applying quantum mechanical perturbation theory is explained. Specific mathematical results required for a proper understanding of the main text are collected together in four appendixes.

Random walk and translational Brownian motion theories are used in the study of relaxation by intermolecular dipolar interactions. Relaxation by intramolecular interactions is investigated by employing rotational Brownian motion theory. On account of the relatively small values of nuclear magnetic moments a one-molecule theory is adequate. The calculations for intramolecular interactions provide not only rotational diffusion results but also results that take account of the inertial effects of the molecule. While the latter are not required in the present state of nuclear magnetic resonance experimental accuracy, they may be of some consequence in the future. For this reason and also because the mathematical methods are applicable to other relaxation processes, including dielectric relaxation where inertial effects can be very important, results are given for both inertial and rotational diffusion theory.

It is hoped that physicists, physical chemists and others interested in nuclear magnetic resonance may find in this book a readable but not superficial account of the derivation of the main results in relaxation theory. For the applied mathematician the book may serve as an illustration of the application of stochastic processes. The appendixes, Chapters 2 and 3, and parts of later chapters could be of assistance to theorists whose interests are outside nuclear magnetic resonance.

My gratitude is due to A. Abragam, E. Belorizky, N. Bloembergen, J. F. Harmon, A. Rigamonti, H. C. Torrey and M. D. Zeidler for allowing me to quote from their published work. I also wish to thank the following for permission to use published material: American Institute of Physics, The American Physical Society, *Berichte der Bunsengesellschaft für Physikalische Chemie*, Clarendon Press, Elsevier Science Publishers, *Journal de Physique*, Taylor and Francis, Ltd. I am indebted to Miss E. R. Wills for invaluable assistance in providing bibliographical material and to Mrs E. Maguire for painstakingly typing the manuscript.

Dublin   *James McConnell*
May, 1986

1

Phenomenological theory of relaxation

1.1 Nuclear magnetic relaxation

In order to establish a microscopic theory of nuclear magnetic relaxation we first study the behaviour of an atomic nucleus in the presence of a constant magnetic field H_0 in a fixed direction, which we take to be the positive z-direction. If the spin vector of the nucleus is the operator $\mathbf{I}$, the nuclear angular momentum operator $\mathbf{m}$ is expressed in terms of $\mathbf{I}$ by (McConnell, 1960; Schiff, 1968)

$$\mathbf{m} = \hbar\mathbf{I}, \tag{1.1}$$

where $\hbar$ is the Planck constant divided by 2π. The measured values of I_z are the eigenvalues of I_z, and these eigenvalues are numbers:

$$-I, -I+1, -I+2, \ldots, I-1, I, \tag{1.2}$$

where I is a positive integer or half-odd integer. We call I the *spin* of the nucleus. Moreover the measured value of the total angular momentum operator $(m_x^2 + m_y^2 + m_z^2)^{1/2}$ is $\{I(I+1)\}^{1/2}\hbar$.

A nucleus with spin angular momentum $\mathbf{m}$ has a permanent spin magnetic moment $\boldsymbol{\mu}$, given by

$$\boldsymbol{\mu} = \gamma\mathbf{m}, \tag{1.3}$$

where γ is the *gyromagnetic ratio* of the nucleus. This is positive, if $\boldsymbol{\mu}$ and $\mathbf{m}$ are parallel, and it is negative if they are antiparallel. A similar phenomenon occurs in the case of the electron and for it the gyromagnetic ratio is almost exactly the *Bohr magneton* $e\hbar/(2m_ec)$, where e is the electronic charge, m_e the mass of the electron and c the velocity light *in vacuo*. If m_e is replaced by the mass m_p of the proton, we have the *nuclear magneton* β_N:

$$\beta_N = \frac{e\hbar}{2m_pc}. \tag{1.4}$$

The nuclear magnetic moment is often expressed by

$$\boldsymbol{\mu} = g\beta_N\mathbf{I}, \tag{1.5}$$

where g is a dimensionless quantity called the *nuclear g factor*. For the proton g is 2.7927 and for the neutron it is -1.913, the negative sign indicating that the magnetic moment is in the direction opposite to that of the spin angular momentum. The values of physical quantities will often be taken from Royal Society (1975). The magnetic moment of the proton is 1.4106×10^{-26} J T^{-1}.

For all nuclei the absolute value of g lies between 0.1 and 6 approximately (Pople, Schneider & Bernstein, 1959). Since $m_p = 1836 m_e$, it follows from (1.4) and (1.5) that the ratio of a nuclear magnetic moment to the electron magnetic moment is of order 10^{-3}–10^{-4}. From (1.1), (1.3) and (1.5) we see that

$$\mu = \gamma \hbar \mathbf{I} = g \beta_N \mathbf{I}, \tag{1.6}$$

so that the magnetic moments of different nuclei are specified by the values of their spins and their g factors.

According to classical electrodynamics a nucleus with magnetic moment μ under the influence of a magnetic field of intensity $\mathbf{H}$ experiences a torque $\mu \times \mathbf{H}$ (McConnell, 1980b, p. 2) and therefore its angular momentum $\mathbf{m}$ satisfies

$$\frac{d\mathbf{m}}{dt} = \mu \times \mathbf{H}.$$

Hence, from (1.3),

$$\frac{d\mu}{dt} = \gamma (\mu \times \mathbf{H}). \tag{1.7}$$

The Hamiltonian $\mathscr{H}$ for the interaction of the nucleus with the field is given by (Jeans, 1933, p. 377)

$$\mathscr{H} = -(\mu_x H_x + \mu_y H_y + \mu_z H_z).$$

On employing the quantum mechanical equations of motion

$$\frac{dm_x}{dt} = \frac{i}{\hbar} (\mathscr{H} m_x - m_x \mathscr{H}),$$

etc., and the commutation relations

$$m_y m_z - m_z m_y = i\hbar m_x,$$

etc. (McConnell, 1960), it is easily seen that (1.7) is valid also in quantum theory.

Suppose that the magnetic field is in the z-direction and that its intensity has the constant value H_0. Then (1.7) yields

$$\frac{d\mu_x}{dt} = \gamma \mu_y H_0, \qquad \frac{d\mu_y}{dt} = -\gamma \mu_x H_0, \qquad \frac{d\mu_z}{dt} = 0. \tag{1.8}$$

Hence

$$\frac{d^2\mu_x}{dt^2} = -\gamma^2 H_0^2 \mu_x, \qquad \frac{d^2\mu_y}{dt^2} = -\gamma^2 H_0^2 \mu_y$$

and by suitably choosing time zero we may express the solution of (1.8) as

$$\mu_x = \mu_\perp \cos \omega_0 t, \qquad \mu_y = -\mu_\perp \sin \omega_0 t, \qquad \mu_z = \mu_\parallel, \qquad (1.9)$$

where

$$\omega_0 = \gamma H_0 \qquad (1.10)$$

and $\mu_\perp$, $\mu_\parallel$ are constants. Equations (1.9) show that the dipole axis of the nucleus precesses about H_0 with angular velocity $-\omega_0$. We call the motion of the dipole *Larmor precession*, ω_0 the *Larmor angular frequency* and v_0 defined by

$$v_0 = \frac{\gamma H_0}{2\pi} \qquad (1.11)$$

the *Larmor frequency*.

The energy of the nucleus in the presence of H_0 is $-\mu_z H_0$, and according to (1.2), (1.6) and (1.10) the energy levels are

$$I\hbar\omega_0, (I-1)\hbar\omega_0, \ldots, -(I-1)\hbar\omega_0, -I\hbar\omega_0. \qquad (1.12)$$

In principle these energy levels could be disturbed by the fields of neighbouring nuclei. When we are interested only in liquids that are in steady state thermal motion, the fields of the other nuclei will average out and we may therefore accept (1.12) as providing the energy levels, if H_0 is sufficiently strong. The energy difference between consecutive levels is $\pm\hbar\omega_0$. Hence in order to raise the nuclear spins from one level to the next highest level we should irradiate the nuclei with electromagnetic waves of frequency $|v_0|$ given by (1.11). Such a process is called *nuclear magnetic resonance absorption*, and it gives rise to a sharp spectral line. In general the study of magnetic resonance is concerned with observing transitions caused by the field whose frequency corresponds to the Larmor precession of the magnetic nuclei around a constant field. This frequency lies in the radiofrequency (rf) range.

Since we are dealing with steady state motion, the populations of nuclei in the various levels obey the Boltzmann distribution law

$$\frac{\exp[-E_{m'}/kT]}{\sum_{m'} \exp[-E_{m'}/kT]}. \qquad (1.13)$$

In the present case the values of the energies $E_{m'}$ are given by (1.12) and let us therefore put

$$E_{m'} = -m'\gamma\hbar H_0, \qquad (1.14)$$

where m' assumes the values (1.2). We now consider the case of identical nuclei, their number per unit volume being N. By symmetry the resultant magnetization is in the z-direction. We obtain the magnetization M per unit volume by averaging the magnetic moment $\gamma \hbar m'$ over the energy states and multiplying by N. Hence, we deduce from (1.13) and (1.14) that

$$M = N \frac{\sum_{m'=-I}^{I} \gamma \hbar m' \exp[\gamma \hbar m' H_0 / kT]}{\sum_{m'=-I}^{I} \exp[\gamma \hbar m' H_0 / kT]}. \qquad (1.15)$$

To estimate the magnitude of the exponent we put m' equal to unity, take H_0 equal to 1 T and the nucleus to be a proton so that (Pople *et al.*, 1959, Appendix A) $\gamma H_0 = 2\pi \times 4.258 \times 10^7 \, \text{s}^{-1}$, put $\hbar = 1.055 \times 10^{-34} \, \text{J s}$ and $k = 1.381 \times 10^{-23} \, \text{J K}^{-1}$, thus obtaining for room temperature (300 K)

$$\frac{\gamma \hbar m' H_0}{kT} = \frac{\gamma \hbar H_0}{kT} = 6.813 \times 10^{-6}, \qquad (1.16)$$

which is very much less than unity. On expanding the exponentials in (1.15) we obtain approximately

$$M = \frac{N\gamma \hbar \sum_{m'=-I}^{I} m'\left(1 + \frac{\gamma \hbar m' H_0}{kT}\right)}{\sum_{m'=-I}^{I}\left(1 + \frac{\gamma \hbar m' H_0}{kT}\right)}.$$

Employing the results

$$\sum_{m'=-I}^{I} 1 = 2I + 1, \qquad \sum_{m'=-I}^{I} m' = 0,$$

$$\sum_{m'=-I}^{I} m'^2 = \tfrac{1}{3}I(I+1)(2I+1)$$

we find that

$$M = \frac{N\gamma^2 \hbar^2 I(I+1)}{3kT} H_0. \qquad (1.17)$$

The multiplier of H_0 is the *static nuclear susceptibility*, which we write χ_0, so that

$$\chi_0 = \frac{N\gamma^2 \hbar^2 I(I+1)}{3kT}. \qquad (1.18)$$

The susceptibility is a macroscopic quantity which is expressed in terms of microscopic quantities by (1.18). Since, from (1.6)

$$\mu^2 = \gamma^2 \hbar^2 I(I+1),$$

(1.18) is equivalent to

$$\chi_0 = \frac{N\mu^2}{3kT}. \tag{1.19}$$

This is Curie's law (Curie, 1895), which was established theoretically by Langevin (1905). We see from (1.19) that the susceptibility is positive and therefore, by definition, the collection of nuclei is *paramagnetic*. It is moreover temperature dependent. The decrease in χ_0 resulting from the increase in temperature is due to the increased randomization of the orientations of the dipoles. We also note that, on account of the μ^2-factor in (1.19), the nuclear susceptibilities are of order 10^{-6} to 10^{-8} that of the electron paramagnetic susceptibility.

When there is no external field, $E_{m'}$ vanishes by (1.14). Then the energy levels have equal populations for steady state motion. When the external field H_0 is effective, (1.13) shows that, when a steady state has been attained, the populations of the various energy levels are different. This means that the application of the external field produces changes in the spin orientations, and this in turn produces an increase $-(\mathbf{M}\cdot\mathbf{H}_0)$ of energy per unit volume in the spin system. By (1.17) this quantity is negative and the surplus energy must be dispersed throughout the environment composed of the molecules which constitute the thermal motion. This environment is called the *lattice*; this term is not confined to atoms in a crystal lattice but is applied also to a liquid or a gaseous medium.

We shall now consider some effects of the mutual interactions of identical nuclear spins on each other. If there is no external magnetic field, the ensemble of spins will be in thermal equilibrium, there will be no preferential direction for the spins, and hence $\mathbf{M}$ will be zero. If a constant field $\mathbf{H}_0$ in a fixed direction is applied, this will produce for each spin with magnetic moment $\boldsymbol{\mu}$ an interaction energy $-(\boldsymbol{\mu}\cdot\mathbf{H}_0)$. When thermal equilibrium is attained, the spins will have a Boltzmann distribution deduced from (1.13) by putting $E_{m'} = -(\boldsymbol{\mu}\cdot\mathbf{H}_0)$ and summing over the spins. Thus the spins will be preferentially in states where $(\boldsymbol{\mu}\cdot\mathbf{H}_0)$ has large values, and consequently $\mathbf{M}$ will be in the direction of $\mathbf{H}_0$.

If the field is changed to another fixed value $\mathbf{H}_0'$ in the same direction, the system is disturbed, the orientations of the dipoles will change and the new magnetization $\mathbf{M}'$ per unit volume will have a component $M_{\parallel}'$ in the direction of $\mathbf{H}_0'$ and a component $M_{\perp}'$ in a transverse direction. When the system reaches a new steady state of equilibrium, the components will satisfy

$$M_{\parallel}' = M', \qquad M_{\perp}' = 0,$$

where

$$M' = \chi_0 H'_0$$

and χ_0 is given by (1.18). The approach to such a state is called *nuclear magnetic relaxation*. The approach of $M'_\parallel$ to its equilibrium value is called *longitudinal relaxation* and the approach of $M'_\perp$ to zero is called *transverse relaxation*.

In the nuclear magnetic relaxation processes that we shall investigate a system of particles having spins and magnetic moments is made to interact with a strong constant field in the z-direction and also with a weak time dependent perturbing field, the interaction Hamiltonian involving the particle spins. If M_0 is the value of M_z when equilibrium has been reached, it is frequently found that M_z obeys an equation

$$\frac{dM_z}{dt} = -\frac{M_z - M_0}{T_1}. \tag{1.20}$$

The longitudinal relaxation is caused by the interaction of the spins with the lattice and is therefore often called *spin–lattice relaxation*. T_1 is the *spin–lattice relaxation time* or *longitudinal relaxation time*. Equation (1.20) shows that $M_z - M_0$ tends to zero with e^{-t/T_1}.

While the motion is settling down to its steady state, the individual magnetic particles precess about the z-axis. This precessional motion is influenced by the internal field arising from interactions with spins of neighbouring particles. This internal field does not contribute to the total energy of the system. However, it has the effect that the particles do not all precess with the same angular velocity, and so the transverse components M_x, M_y tend to zero. If there exists an equation

$$\frac{dM_x}{dt} = -\frac{M_x}{T_2} \tag{1.21}$$

and we are dealing with isotropic media, there will also be an equation

$$\frac{dM_y}{dt} = -\frac{M_y}{T_2}. \tag{1.22}$$

T_2 is called the *spin–spin relaxation time* or *transverse relaxation time*. Similarly transverse relaxation is called *spin–spin relaxation*. For solids it it usually found that $T_1 \gg T_2$, whereas for liquids $T_1 \approx T_2$. The quantities T_1^{-1}, T_2^{-1} are called *relaxation rates*.

1.2 The Bloch equations

We consider the time variations of the components of **M**, the magnetization per unit volume resulting from a constant magnetic field **H** in a fixed direction. From (1.7) we deduce that

$$\frac{\mathrm{d}M_x}{\mathrm{d}t} = \gamma(M_y H_z - M_z H_y),$$

$$\frac{\mathrm{d}M_y}{\mathrm{d}t} = \gamma(M_z H_x - M_x H_z), \qquad (1.23)$$

$$\frac{\mathrm{d}M_z}{\mathrm{d}t} = \gamma(M_x H_y - M_y H_x),$$

provided that the interactions of the spins between themselves and with their environment are neglected. Bloch (1946) made two assumptions:

(a) In order to include the influence of the neglected interactions we combine (1.20)–(1.22) with (1.23).

(b) H_x, H_y, H_z need not be constants.

He thus proposed the *Bloch equations*:

$$\frac{\mathrm{d}M_x}{\mathrm{d}t} - \gamma(M_y H_z - M_z H_y) + \frac{M_x}{T_2} = 0,$$

$$\frac{\mathrm{d}M_y}{\mathrm{d}t} - \gamma(M_z H_x - M_x H_z) + \frac{M_y}{T_2} = 0, \qquad (1.24)$$

$$\frac{\mathrm{d}M_z}{\mathrm{d}t} - \gamma(M_x H_y - M_y H_x) + \frac{M_z - M_0}{T_1} = 0.$$

These equations are phenomenological but are nevertheless very useful for the study of nuclear induction.

Bloch considered the case of

$$H_x = H_1 \cos \omega t, \qquad H_y = -H_1 \sin \omega t, \qquad H_z = H_0 \qquad (1.25)$$

corresponding to a constant rotating rf field H_1 perpendicular to a constant field H_0 and rotating about it in a clockwise direction. He supposed that both H_1 and H_0 are positive and that $H_1 \ll H_0$. On substituting (1.25) into (1.24) we have

$$\frac{\mathrm{d}M_x}{\mathrm{d}t} - \gamma(M_y H_0 + M_z H_1 \sin \omega t) + \frac{M_x}{T_2} = 0,$$

$$\frac{\mathrm{d}M_y}{\mathrm{d}t} - \gamma(M_z H_1 \cos \omega t - M_x H_0) + \frac{M_y}{T_2} = 0, \qquad (1.26)$$

$$\frac{\mathrm{d}M_z}{\mathrm{d}t} + \gamma(M_x H_1 \sin \omega t + M_y H_1 \cos \omega t) + \frac{M_z - M_0}{T_1} = 0.$$

To discuss (1.26) we transform to a new cartesian coordinate system shown in Fig. 1.1 such that the z'-axis coincides with z-axis, the x'-axis is in the direction of $\mathbf{H}_1$ and consequently the y'-axis makes the angle ωt with the y-axis. We denote by u and v the components of $\mathbf{M}$ in the x'- and y'-directions, respectively, so that

$$M_x = u\cos\omega t + v\sin\omega t, \qquad M_y = -u\sin\omega t + v\cos\omega t,$$
$$u = M_x\cos\omega t - M_y\sin\omega t, \qquad v = M_x\sin\omega t + M_y\cos\omega t. \qquad (1.27)$$

On substitution from (1.27) into (1.26) we deduce that

$$\frac{du}{dt} - (\omega_0 - \omega)v + \frac{u}{T_2} = 0,$$

$$\frac{dv}{dt} + (\omega_0 - \omega)u - \gamma H_1 M_z + \frac{v}{T_2} = 0, \qquad (1.28)$$

$$\frac{dM_z}{dt} + \gamma H_1 v + \frac{M_z - M_0}{T_1} = 0,$$

where γH_0 is written ω_0 according to (1.10). If in (1.24) we make the substitutions

$$M_x \mapsto u, \qquad M_y \mapsto v, \qquad M_z \mapsto M_z$$
$$H_x \mapsto H_1, \qquad H_y \mapsto 0, \qquad H_z \mapsto H_0 - \frac{\omega}{\gamma},$$

Fig. 1.1 The coordinate system with x'-axis in the direction of the constant rotating rf field H_1. The positive third axis is perpendicular to the plane of the paper and upwards.

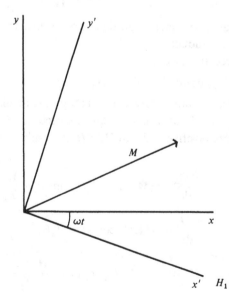

we obtain (1.28). Thus the Bloch equations hold in the rotating frame for the *effective field* $\mathbf{H}_e$ given by

$$\mathbf{H}_e = H_1 \mathbf{i} + \left(H_0 - \frac{\omega}{\gamma} \right) \mathbf{k}, \tag{1.29}$$

where $\mathbf{i}, \mathbf{j}, \mathbf{k}$ are unit vectors in the directions of $0x'$, $0y'$, $0z'$.

The spin–lattice relaxation time in the rotating frame $T_{1\rho}$ is defined as the time of decay of the component of magnetization in the direction of H_e. According to (1.29) this direction makes with $0z'$ an angle θ such that

$$\tan \theta = \frac{\gamma H_1}{\omega_0 - \omega}.$$

The spin–spin relaxation time in the rotating frame is denoted by $T_{2\rho}$. An expression for $T_{1\rho}$ in the case of weak collisions was derived by Jones (1966).

The steady state solutions of (1.28) are found by putting the time derivatives equal to zero, so that we have

$$u = (\omega_0 - \omega) v T_2,$$
$$v = -(\omega_0 - \omega) u T_2 + \gamma H_1 M_z T_2,$$
$$M_z = M_0 - \gamma H_1 T_1 v.$$

It is easily found that

$$\frac{u}{M_0} = \frac{\gamma H_1 (\omega_0 - \omega) T_2^2}{1 + (\omega_0 - \omega)^2 T_2^2 + \gamma^2 H_1^2 T_1 T_2},$$

$$\frac{v}{M_0} = \frac{\gamma H_1 T_2}{1 + (\omega_0 - \omega)^2 T_2^2 + \gamma^2 H_1^2 T_1 T_2}, \tag{1.30}$$

$$\frac{M_z}{M_0} = \frac{1 + (\omega_0 - \omega)^2 T_2^2}{1 + (\omega_0 - \omega)^2 T_2^2 + \gamma^2 H_1^2 T_1 T_2}.$$

It follows from (1.27) that

$$\frac{M_x}{M_0} = \frac{\gamma H_1 T_2 \{ (\omega_0 - \omega) T_2 \cos \omega t + \sin \omega t \}}{1 + (\omega_0 - \omega)^2 T_2^2 + \gamma^2 H_1^2 T_1 T_2},$$

$$\frac{M_y}{M_0} = \frac{\gamma H_1 T_2 \{ -(\omega_0 - \omega) T_2 \sin \omega t + \cos \omega t \}}{1 + (\omega_0 - \omega)^2 T_2^2 + \gamma^2 H_1^2 T_1 T_2}. \tag{1.31}$$

We use these results to examine the case of a constant external field H_0 in the z-direction and a periodic rf field $2H_1 \cos \omega t$ in the x-direction. This system may be replaced by the H_0 and H_1 fields defined in (1.25) together with an H_1 field rotating with angular velocity $-\omega$. Then, since we are employing a linear theory, we obtain from (1.31)

$$\frac{M_x}{M_0} = \gamma H_1 T_2 \left\{ \frac{(\omega_0 - \omega)T_2 \cos \omega t + \sin \omega t}{1 + (\omega_0 - \omega)^2 T_2^2 + \gamma^2 H_1^2 T_1 T_2} + \frac{(\omega_0 + \omega)T_2 \cos \omega t - \sin \omega t}{1 + (\omega_0 + \omega)^2 T_2^2 + \gamma^2 H_1^2 T_1 T_2} \right\}.$$

(1.32)

Let us now consider resonance effects due to a weak periodic field $2H_1 \cos \omega t$. Since the Larmor precession of the spins is clockwise, resonance can arise only from the component of $2H_1 \cos \omega t$ which results from the rotation in the clockwise direction. Hence we neglect the second term in the braces of (1.32), and M_x is obtained from the first equation of (1.31). Introducing from (1.10), (1.17) and (1.18)

$$\gamma H_0 = \omega_0, \qquad M_0 = \chi_0 H_0,$$

(1.33)

where χ_0 is the static nuclear susceptibility, we have

$$M_x = \frac{\chi_0 \omega_0 H_1 T_2 \{(\omega_0 - \omega)T_2 \cos \omega t + \sin \omega t\}}{1 + (\omega_0 - \omega)^2 T_2^2 + \gamma^2 H_1^2 T_1 T_2}.$$

(1.34)

Then writing

$$M_x = 2H_1 \cos \omega t \, \chi'(\omega) + 2H_1 \sin \omega t \, \chi''(\omega)$$

(1.35)

we see from (1.34) and (1.35) that

$$\chi'(\omega) = \frac{\frac{1}{2}\chi_0 \omega_0 (\omega_0 - \omega)T_2^2}{1 + (\omega_0 - \omega)^2 T_2^2 + \gamma^2 H_1^2 T_1 T_2},$$

(1.36)

$$\chi''(\omega) = \frac{\frac{1}{2}\chi_0 \omega_0 T_2}{1 + (\omega_0 - \omega)^2 T_2^2 + \gamma^2 H_1^2 T_1 T_2}.$$

(1.37)

If we put

$$\chi(\omega) = \chi'(\omega) + i\chi''(\omega),$$

(1.38)

where we call $\chi(\omega)$ the *complex magnetic susceptibility*, then (1.35) may be expressed as

$$M_x = \mathrm{Re}(2H_1 \chi(\omega) \, \mathrm{e}^{-i\omega t}),$$

where Re denotes (real part of).

We return to the rotating system of Fig. 1.1 in order to derive an expression for the time rate of absorption of energy per unit volume. Employing the expression $\mu \times H_1$ for the torque exerted on a magnetic moment μ by H_1 we see from (1.27) that H_1 produces on the nuclear spin system a torque of moment $-vH_1$ per unit volume about the positive z-axis. If $\phi = -\omega t$, the angle through which the x'- and y'-axes have turned about the z-axis, the work done per unit volume by H_1 is $-vH_1\phi$. Since we are concerned with steady motion, dv/dt vanishes and the rate of work done is $vH_1\omega$; in other words, the power absorbed by the system is $vH_1\omega$. From (1.30), (1.33) and (1.37) the *power absorption* $b(\omega)$ is given by

$$b(\omega) \equiv vH_1\omega = \frac{\omega\omega_0\chi_0 H_1^2 T_2}{1+(\omega_0-\omega)^2 T_2^2 + \gamma^2 H_1^2 T_1 T_2} \tag{1.39}$$

and

$$b(\omega) = 2\omega H_1^2 \chi''(\omega). \tag{1.40}$$

If we take $|\omega_0-\omega| \ll \omega_0$, we may approximate (1.39) and (1.40) by

$$b(\omega) = \frac{\omega_0^2 \chi_0 H_1^2 T_2}{1+(\omega_0-\omega)^2 T_2^2 + \gamma^2 H_1^2 T_1 T_2}, \tag{1.41}$$

$$b(\omega) = 2\omega_0 H_1^2 \chi''(\omega), \tag{1.42}$$

so that $b(\omega)$ is proportional to $\chi''(\omega)$.

When $\gamma^2 H_1^2 T_1 T_2 \ll 1$, (1.36) and (1.37) become

$$\chi'(\omega) = \frac{\tfrac{1}{2}\chi_0 \omega_0 (\omega_0-\omega) T_2^2}{1+(\omega_0-\omega)^2 T_2^2} \tag{1.43}$$

$$\chi''(\omega) = \frac{\tfrac{1}{2}\chi_0 \omega_0 T_2}{1+(\omega_0-\omega)^2 T_2^2}. \tag{1.44}$$

If we put $(\omega-\omega_0)T_2 = z$, we see that $\chi''(\omega)$ is proportional to $(1+z^2)^{-1}$. The graph of this as a function of z is called a *Lorentzian curve*. It is symmetric about $z=0$, its maximum value is unity and it decreases monotonically to zero as $z \to \pm\infty$. The value $\tfrac{1}{2}$ of the function occurs for $z = \pm 1$ and the breadth at half-maximum is therefore 2. The graph of $\chi''(\omega)$ as a function of ω given by (1.44) and shown in Fig. 1.2 is a Lorentzian curve with maximum value $\tfrac{1}{2}\chi_0 \omega_0 T_2$, the breadth at half-maximum being $2T_2^{-1}$. We see from (1.43) that $\chi'(\omega)$ is an odd function of $(\omega-\omega_0)T_2$. Its maximum

Fig. 1.2 Graph of $\chi''(\omega)$ as a function of ω.

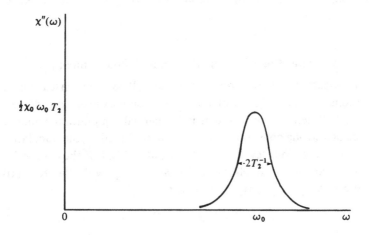

and minimum values $\pm\frac{1}{4}\chi_0\omega_0 T_2$ are attained at $\omega-\omega_0=\mp T_2^{-1}$, and the function tends to zero as $\omega-\omega_0$ tends to $\pm\infty$.

Another consequence of the condition $\gamma^2 H_1^2 T_1 T_2 \ll 1$ is, as we see from (1.41), that $b(\omega)$ becomes proportional to H_1^2. Thus the power absorption may be increased by increasing the amplitude $2H_1$ of the alternating rf field.

Let us now examine whether it is reasonable to consider the condition $\gamma^2 H_1^2 T_1 T_2 \gg 1$ in view of the restriction $H_1 \ll H_0$ that Bloch placed on the rf field. The values of T_1 and T_2 for liquids are nearly equal and their common value is in the range 10^{-4} to 10^2 s (Bloembergen et al., 1948). Hence the condition $\gamma H_1 (T_1 T_2)^{1/2} \gg 1$ is equivalent to $\gamma H_1 \gg 10^4$ s^{-1}. From the figures given in section 1.1 the value of γH_0 for a proton in a magnetic field of 1 T is 2.675×10^8 s^{-1}. It will therefore be easy to satisfy both $H_1 \ll H_0$ and $\gamma^2 H_1^2 T_1 T_2 \gg 1$. Let us, for example, take $\gamma H_1 = 10^6$ s^{-1}. Then we obtain from (1.37)

$$\chi''(\omega)=\frac{\frac{1}{2}\chi_0\omega_0}{\gamma^2 H_1^2 T_1\left[1+\frac{(\omega_0-\omega)^2}{\gamma^2 H_1^2}\frac{T_2}{T_1}\right]}. \qquad (1.45)$$

Since $T_2 \approx T_1$, the last term in the bracket may be neglected for values of ω such that $|\omega_0-\omega| \ll 10^6$ s^{-1} and (1.45) reduces to

$$\chi''(\omega)=\frac{\chi_0\omega_0}{2\gamma^2 H_1^2 T_1}. \qquad (1.46)$$

Then, since $|\omega_0-\omega| \ll \omega_0 \approx 10^8$ s^{-1}, we deduce from (1.42) and (1.46) that

$$b(\omega)=\frac{\omega_0^2\chi_0}{\gamma^2 T_1},$$

which is independent of H_1. This tendency towards a constant power absorption is called *saturation* and $(1+\gamma^2 H_1^2 T_1 T_2)^{-1}$ is called the *saturation factor*.

1.3 The Bloembergen, Purcell and Pound theory

The investigations of Bloembergen et al. (1948) are based on the populations of energy levels of the nuclear spin system considered in section 1.1. When the spin system is in thermal equilibrium under the influence of a strong constant field H_0 in a fixed direction, the populations of nuclei in the various levels obey (1.13) and (1.14). For the case of $I=\frac{1}{2}$ there are two energy levels given by $m'=\frac{1}{2}, -\frac{1}{2}$, with the respective populations $N_{1/2}, N_{-1/2}$, say. Let us write

$$n=N_{1/2}-N_{-1/2}, \qquad (1.47)$$

the surplus population. On account of the different Boltzmann density factors, n will not vanish. We denote by n_0 the value of n corresponding to thermal equilibrium at the lattice temperature. Bloembergen *et al.* made the assumption that in the absence of the rf field introduced in the previous section n obeys an equation

$$\frac{dn}{dt} = -\frac{n - n_0}{T_1},$$ (1.48)

where T_1 is the spin–lattice relaxation time. In the presence of the field H_0 in the z-direction and a weak oscillating field $2H_1 \, e^{i\omega t}$ in the x-direction they modified (1.48) to

$$\frac{dn}{dt} = -\frac{n - n_0}{T_1} - 2nW,$$ (1.49)

where W is the probability of a single transition in which the I_z of a nucleus changes from $\tfrac{1}{2}$ to $-\tfrac{1}{2}$. The value of W is expressed by

$$W = \tfrac{1}{4}\gamma^2 H_1^2 g(v),$$ (1.50)

where $g(v)$ gives the observed shape of the absorbed line, $g(v)$ being normalized so as to satisfy

$$\int_0^\infty g(v) \, dv = 1,$$

v being the frequency within the width of the levels from which radiation occurs. Then, from (1.49) and (1.50),

$$\frac{dn}{dt} = -\frac{n - n_0}{T_1} - \frac{n}{2}\gamma^2 H_1^2 g(v).$$ (1.51)

If n_s is the steady state value of n, (1.51) gives

$$\frac{n_s}{n_0} = \frac{1}{1 + \tfrac{1}{2}\gamma^2 H_1^2 T_1 g(v)}.$$

When a nucleus of spin $\tfrac{1}{2}$ interacts with the constant field H_0, the interaction energy is $-\tfrac{1}{2}\gamma h H_0$ for $I_z = \tfrac{1}{2}$ and $\tfrac{1}{2}\gamma h H_0$ for $I_z = -\tfrac{1}{2}$. Let us for simplicity take γ to be positive; if it is negative, we just change the roles of $I_z = \tfrac{1}{2}$ and $I_z = -\tfrac{1}{2}$ in the present section. Consider now the case of thermal equilibrium of spin $\tfrac{1}{2}$ nuclei in interaction with a lattice of temperature T. If $p_{1/2}, p_{-1/2}$ are the probabilities for $I_z = \tfrac{1}{2}, -\tfrac{1}{2}$ nuclei, then from the Boltzmann distribution law (1.13)

$$\frac{p_{1/2}}{p_{-1/2}} = \exp\left(\frac{\gamma h H_0}{kT}\right).$$ (1.52)

This shows that $p_{1/2} > p_{-1/2}$ and hence that $N_{1/2} > N_{-1/2}$ for thermal equilibrium with the lattice. Thus n in (1.47) is a positive integer.

Let us now consider the question of thermal equilibrium in the presence of both spin–lattice and spin–spin interactions. When a rf field is applied, transitions between the two spin states are induced. According to the principle of detailed balance (Heitler, 1954) the probabilities of a single transition in which the I_z of a nucleus changes from $-\frac{1}{2}$ to $\frac{1}{2}$ is also W. Hence $N_{1/2}W$ nuclei change their I_z from $\frac{1}{2}$ to $-\frac{1}{2}$ and $N_{-1/2}W$ change their I_z from $-\frac{1}{2}$ to $\frac{1}{2}$. Since $N_{1/2} > N_{-1/2}$, the population of $I_z = \frac{1}{2}$ nuclei decreases and the population of $I_z = -\frac{1}{2}$ nuclei increases. Consequently, if we denote by $p'_{1/2}, p'_{-1/2}$ the new probabilities for $I_z = \frac{1}{2}, -\frac{1}{2}$ nuclei,

$$\frac{p'_{1/2}}{p'_{-1/2}} < \frac{p_{1/2}}{p_{-1/2}}. \tag{1.53}$$

Just as *lattice temperature* T may be defined by (1.52) we now define the *spin temperature* T_s by

$$\frac{p'_{1/2}}{p'_{-1/2}} = \exp\left(\frac{\gamma \hbar H_0}{kT_s}\right). \tag{1.54}$$

It follows from (1.52)–(1.54) that $T_s > T$; the spin temperature is greater than the lattice temperature.

We write $N'_{1/2}, N'_{-1/2}$ for the number of nuclei with $I_z = \frac{1}{2}, -\frac{1}{2}$ in the steady state system where both spin–lattice and spin–spin interactions are operative. We then put

$$n'_s = N'_{1/2} - N'_{-1/2} \tag{1.55}$$

in analogy with (1.47). Since $N'_{1/2}, N'_{-1/2}$ are proportional to $p'_{1/2}, p'_{-1/2}$, we deduce from (1.54) and (1.55) that n'_s vanishes when $kT_s \gg \gamma \hbar H_0$. When n'_s vanishes, we say that saturation occurs. There is then equilibrium between the spin system and the lattice, the rf field supplying energy at a constant rate to the spin system in order to maintain the spin temperature T_s at its value above the temperature T of the environment. Thus there is a constant power absorption and the above description of saturation is consistent with that given at the end of the previous section. When $I > \frac{1}{2}$, the definition of spin temperature poses certain difficulties (Abragam, 1961, p. 134).

Bloembergen *et al.* derived expressions for relaxation times. By analyzing the dipole–dipole interaction between two nuclei they obtained for nuclei of spin I

$$\frac{1}{T_1} = \frac{3}{2}\gamma^4 \hbar^2 I(I+1)[J_1(\nu_0) + J_2(2\nu_0)], \tag{1.56}$$

where ν_0 is given by (1.11) and we have omitted an erroneous $\frac{1}{2}$ that appears before $J_2(2\nu_0)$ in their equation (34). The functions J_1, J_2 were calculated by the Debye theory of rotational Brownian motion:

$$J_1(v) = \frac{4}{15r^6} \frac{\tau_c}{1 + 4\pi^2 v^2 \tau_c^2}, \qquad J_2(v) = 4J_1(v), \tag{1.57}$$

where τ_c is a time characteristic of the random motion and r is the distance between two nuclei in the same molecule.

The theories of Bloch and of Bloembergen *et al.* are found to give satisfactory agreement with experiment for nuclear resonances in liquids, in gases, and also in solids when the intensity of the rf field is low. When the intensity is not low, disagreement occurs (Redfield, 1955).

1.4 The different relaxation processes

It is the purpose of this book to present in some detail a microscopic theory of nuclear magnetic processes that give rise to relaxation phenomena in liquids. For nuclei with spin $\frac{1}{2}$ an important process is dipole–dipole interaction. This is an *intramolecular interaction* when the dipoles are in the same molecule, and it is an *intermolecular interaction* when the dipoles are in different molecules. All other interactions that we shall examine are intramolecular. When the spin of a nucleus in a molecule is greater than $\frac{1}{2}$, the nucleus may have a quadrupole moment and the interaction of this moment with the field produced at the nucleus by the other molecular charges yields a relaxation usually more important than that due to the above dipolar interaction. Relaxation by anisotropic chemical shift is caused by the interaction of the nuclear spin with the field arising from the perturbation of shells of molecular electrons by an external field. Similarly when a molecule rotates, the motion of its electrons produces a magnetic field at the nucleus which causes spin-rotational relaxation. Finally in order to supplement our studies of dipolar interactions we shall consider relaxation by a scalar interaction.

In the next two chapters we endeavour to express certain basic mathematical and physical results in a form suitable for the theoretical study of nuclear magnetic relaxation. When recalling mathematical results we leave some material to appendixes so as not to retard unduly the sequence of the main text. There follows a lengthy chapter on dipolar interactions, both intermolecular and intramolecular, and formulae are derived which express the relaxation times T_1 and T_2 in terms of spectral densities. In Chapters 5 and 6 explicit expressions are deduced for T_1 and T_2 for a variety of molecular models. This is complemented by a short chapter on relaxation by scalar interaction. In Chapters 8 and 9 relaxation by anisotropic chemical shift and by quadrupolar interaction are studied and it is found that, on account of certain similarities in the interaction

Hamiltonians, a number of results from Chapter 6 may be employed. This is no longer true in Chapter 10, which deals with relaxation by spin-rotational interactions and which requires a more extended mathematical treatment. Finally the relevance of the theoretical results for nuclear magnetic relaxation experiments is illustrated in Chapter 11.

2

Random motion

2.1 Probability distributions

In the course of the previous chapter we have been concerned with liquids that are in a steady state of thermal motion. Since thermal motion is random, the motion of the molecules in the liquid and consequently the motion of the nuclear spins is random. Thus we are led to the study of random motion. As a preparation for this we first consider probability distributions of a random variable.

A *random variable*, also called a *stochastic variable*, is one whose values are governed by some probability law. The variable may be continuous or discrete. An obvious example of a discrete variable is the face of a die, which is specified by the number of dots on it. Examples of continuous variables are the position and orientation of a molecule of liquid in thermal motion.

Let X be a continuous stochastic variable and for convenience let us suppose that the range of X is one-dimensional. The extension to more than one dimension poses no problem. We denote by $P[X \leqslant x']$ the probability that the value of X be less than or equal to x'. Then it is possible to write

$$P[X \leqslant x'] = \int_{-\infty}^{x'} p(X; x)\, dx, \qquad (2.1)$$

and $p(X; x)$ is defined as the *probability density function*. A very important continuous random variable is the *Gaussian variable*, which may be defined by the probability density function of the Gaussian distribution given by (Jeffreys, 1967, pp. 68–71)

$$p(X; x) = \frac{\exp\left\{\dfrac{(x-m)^2}{2\sigma^2}\right\}}{(2\pi\sigma^2)^{1/2}}. \qquad (2.2)$$

In place of a probability density function a discrete random variable has a *probability mass function* $p(X; x)$ such that the probability of X being found in a subset of all the values of X is $\Sigma p(X; x)$ summed over the subset.

As an example we may take

$$p(X;x)=\frac{e^{-m}m^x}{x!} \qquad (2.3)$$

for the probability mass function of the *Poisson distribution*. Equation (2.3) gives the probability of x events per unit time, if on the average a large number m of events occur per unit time.

It is appropriate at this stage to say something about the notion of averaging. When we speak of an average, this will usually signify the *ensemble average* of a physical system. By this is meant the average at one instant of time over the system under consideration together with a large number of copies of it with which it is in thermal equilibrium. We denote ensemble average by angular brackets, so that for the function $f(X)$ of a continuous variable it is possible to write

$$\langle f(X)\rangle = \int_{-\infty}^{\infty} f(x)p(X;x)\,dx, \qquad (2.4)$$

and for a function of a discrete variable

$$\langle f(X)\rangle = \sum_x f(x)p(X;x). \qquad (2.5)$$

If $\langle f(X)\rangle$ vanishes, $f(X)$ is said to be a *centred random variable*. On substituting from (2.2) into (2.4) we may deduce that (McConnell, 1980b, pp. 45, 46)

$$\langle X\rangle = m, \qquad \langle X^2\rangle = \sigma^2 + m^2.$$

Hence the *variance* $V(X)$, which is general is defined by

$$V(X) = \langle X^2\rangle - \langle X\rangle^2,$$

is equal to σ^2. Similarly we may deduce from (2.3) and (2.5) that for the Poisson distribution

$$\langle X\rangle = m, \qquad V(X) = m.$$

Another type of average is the *time average* which is found by recording many values of the variable, $X(t)$ say, during an interval τ, calculating their average and then allowing τ to be indefinitely long. Denoting the time average by an overhead bar we have

$$\overline{X(t)} = \lim_{\tau\to\infty} \frac{1}{\tau} \int_{-\tau/2}^{\tau/2} x(t)\,dt, \qquad (2.6)$$

whereas $\langle X(t)\rangle$ is obtained by substituting $X(t)$ for $f(X)$ in (2.4) or (2.5). The *ergodic theorem* states that, when one is dealing with a steady state dynamical system that is under the influence of a static field, which may be

zero,
$$\bar{X}(t)=\langle X(t)\rangle; \tag{2.7}$$

that is, the time average is equal to the ensemble average (Landau & Lifschitz, 1958, section 1). Thus the ergodic theorem is applicable to a system of nuclear spins that are being tossed around by the thermal motion of their environment.

We have so far considered a single stochastic variable X, whose range however need not be restricted to one dimension. We now consider more than one X; for simplicity of exposition we take only two variables X_1 and X_2, which we suppose to be continuous. Then the probability that the value of X_1 be less than or equal to x_1' and that the value of X_2 be less than or equal to x_2' is expressed as

$$\int_{-\infty}^{x_1'} dx_1 \int_{-\infty}^{x_2'} dx_2\, p(X_1,x_1;X_2,x_2), \tag{2.8}$$

where $p(X_1,x_1;X_2,x_2)$ is called the *joint probability density function*. When X_1 and X_2 are independent, the double integral will be equal to the product of the probabilities that X_1 is less than or equal to x_1' and that X_2 is less than or equal to x_2'. Hence (2.8) is expressible as

$$\int_{-\infty}^{x_1'} dx_1\, p(X_1;x_1) \int_{-\infty}^{x_2'} dx_2\, p(X_2;x_2)$$

and so
$$p(X_1,x_1;X_2,x_2)=p(X_1;x_1)p(X_2;x_2). \tag{2.9}$$

Extending (2.4) we define the ensemble average of $f(X_1,X_2)$ by

$$\langle f(X_1,X_2)\rangle = \int_{-\infty}^{\infty} dx_1\, f(x_1,x_2)p(X_1,x_1;X_2,x_2).$$

If X_1,X_2 are independent variables and if $f(X_1,X_2)$ is the product $f_1(X_1)f_2(X_2)$, it follows from (2.9) that

$$\langle f_1(X_1)f_2(X_2)\rangle = \langle f_1(X_1)\rangle\langle f_2(X_2)\rangle. \tag{2.10}$$

An important concept is that of conditional probability density which enables us to link $p(X_1,x_1;X_2,x_2)$ with $p(X_1,x_1)$. Suppose that, if $X_1=x_1$, the probability of finding X_2 in (x_2,x_2+dx_2) is $w(X_1,x_1;X_2,x_2)\,dx_2$. Then w is called the *conditional probability density*. Since the probability that X_1 be in (x_1, x_1+dx_1) and that X_2 be in (x_2,x_2+dx_2) is expressible as $p(X_1,x_1)\,dx_1\,w(X_1,x_1;X_2,x_2)\,dx_2$ or as $p(X_1,x_1;X_2,x_2)\,dx_1\,dx_2$, we conclude that

$$p(X_1,x_1;X_2,x_2)=p(X_1,x_1)w(X_1,x_1;X_2x_2). \tag{2.11}$$

If X_1 and X_2 are independent, (2.9) shows that $w(X_1,x_1;X_2,x_2)$ is just $p(X_2;x_2)$, which is otherwise obvious.

2.2 Averaging processes

In our later studies the random variables that we shall encounter are usually functions of time: examples of these are the linear coordinate $x(t)$, the angular coordinate $\theta(t)$, and their respective velocities $v(t)$, $\omega(t)$. Each of them is called a stochastic process. Speaking more generally we define a *random process* or a *stochastic process* as a set of random variables $X(t)$ that is defined over a range of values of t. In order to discuss $X(t)$ we divide the range into a large number of subdivisions at $t_1, t_2, \ldots, t_{n-1}, t_n$ $(t_1 < t_2 \ldots < t_{n-1} < t_n)$ and approximate $X(t)$ by $X(t_1), X(t_2), \ldots, X(t_n)$. We then apply results just obtained for $X_1, X_2, \ldots$ to $X(t_1), X(t_2), \ldots$. Thus we define the conditional probability density $w(X(t_{r-1}), x_{r-1}; X(t_r), x_r)$ by saying that if $X(t_{r-1}) = x_{r-1}$, the probability of finding $X(t_r)$ in $(x_r, x_r + dx_r)$ is $w(X(t_{r-1}), x_{r-1}; X(t_r), x_r)\, dx_r$.

A *Markov process* is defined as a random process such that the probability density function at time t_r is independent of what happened before time t_{r-1}. Hence the probability that $X(t_r)$ be in $(x_r, x_r + dx_r)$ depends only on t_r, x_r and $X(t_{r-1})$. Consequently the only conditional probability density that is relevant at time t_n for a Markov process is $w(X(t_{n-1}), x_{n-1}; X(t_n), x_n)$.

As an illustration of these ideas we may instance the motion of a molecule in a liquid at room temperature. The molecule will suffer impacts from neighbouring particles at the rate of about 10^{21} s^{-1}. Accordingly we can take each interval $t_r - t_{r-1}$ to be macroscopically very small while, on account of the great number of the collisions during the interval, what happens to the molecule after time t_r is unrelated to what happened before time t_{r-1}. Thus any process associated with the random motion of the molecule is a Markov process.

By extending to several variables the function $P[X \leqslant x']$ introduced at the beginning of the preceding section and applying the extended function to random processes we define for any $t_1, t_2, \ldots, t_p$ in the range where $X(t)$ is specified

$$P[X(t_1) \leqslant x'_1, X(t_2) \leqslant x'_x, \ldots, X(t_p) \leqslant x'_p] \qquad (2.12)$$

as the joint probability that the value of $X(t_1)$ is less than or equal to x'_1, the value of $X(t_2)$ is less than or equal to $x'_2, \ldots$, the value of $X(t_p)$ is less than or equal to x'_p. When

$$P[X(t_1 + \tau) \leqslant x'_1, X(t_2 + \tau) \leqslant x'_2, \ldots, X(t_p + \tau) \leqslant x'_p] \qquad (2.13)$$

is equal to (2.12) for all values of $\tau_i + \tau$ in the range for which $X(t)$ is defined, we say that $X(t)$ is a *stationary random process*. This situation arises, for example, when a dynamical system is in thermal motion that has attained a steady state.

Let us take a function $f(X_1(t), X_2(t), \ldots, X_s(t))$ of s stationary random processes $X_1(t), X_2(t), \ldots, X_s(t)$. These processes might be, for example, the canonical variables $q_1(t), \ldots, q_n(t), p_1(t), \ldots, p_n(t)$ of a dynamical system. On account of the equality of (2.12) and (2.13) the stationarity implies that the mean value of f is independent of the time at which it is calculated; we shall take the chosen instant of time to be zero. Thus we shall calculate the *ensemble average* $\langle f(X_1(t), X_2(t), \ldots, X_s(t)) \rangle$ by averaging at time zero over the ensemble which consists of the dynamical system and many copies of it. Taking the average at time zero does not mean that we put $t = 0$ in f; what it means is that we start with different initial values of the set $X_1, X_2, \ldots, X_s$.

2.3 Correlation functions and spectral densities

When calculating relaxation times we shall have to consider time-correlation functions, which we shall call simply correlation functions because we shall not be concerned with space-correlation functions. Let us take two functions $A(s(t))$, $B(s(t))$, where $s(t)$ denotes one or more variables at time t; for example, position or velocity variables or both, or orientational variables, or angular velocity variables. In many of our investigations $s(t)$ will denote merely angular variables like the Euler angles employed to specify the orientation of a rigid body (Synge & Griffith, 1959, p. 259). For brevity we write $A(s(t))$ as $A(t)$ and $B(s(t))$ as $B(t)$. The *correlation function* of A and B is defined for a physical system in a steady state and in the absence of an external field as the ensemble average $\langle A^*(0)B(t) \rangle$, where the star denotes complex conjugate. The ensemble average is given by

$$\langle A^*(0)B(t) \rangle = \int A^*(0)B(t)p(s(0), 0; s(t), t)\, \mathrm{d}s(0)\, \mathrm{d}s(t), \qquad (2.14)$$

where p is the joint probability that the variable s will assume the values $s(0)$ at time zero and $s(t)$ at time t. If we introduce the conditional probability density w by

$$p(s(0), 0; s(t), t) = p(s(0), 0)w(s(0), 0; s(t), t) \qquad (2.15)$$

in analogy with (2.11), we may express (2.14) as

$$\langle A^*(0)B(t) \rangle = \int p(s(0), 0)A^*(0)\, \mathrm{d}s(0) \int w(s(0), 0; s(t), t)B(t)\, \mathrm{d}s(t). \qquad (2.16)$$

Since the system is in a steady state,

$$\langle A^*(t')B(t' + t) \rangle = \langle A^*(0)B(t) \rangle, \qquad (2.17)$$

as we see by putting $X(t) \equiv A^*(0)B(t)$ in (2.12) and (2.13).

If A and B are identical functions, we speak of $\langle A^*(0)B(t)\rangle$ as an *autocorrelation function*; if A and B are different, $\langle A^*(0)B(t)\rangle$ is a *cross-correlation function*. The *normalized autocorrelation function* of $A(t)$ is

$$\frac{\langle A^*(0)A(t)\rangle}{\langle A^*(0)A(0)\rangle}. \tag{2.18}$$

We deduce from (2.17) that

$$\langle A^*(0)A(-t)\rangle = \langle A^*(t)A(0)\rangle.$$

If $A^*(t)$ and $A(0)$ commute, we have that

$$\langle A^*(0)A(-t)\rangle = \langle A^*(0)A(t)\rangle^*. \tag{2.19}$$

Hence, if we also know that the autocorrelation is real, we see from (2.19) that it is an even function of t.

The *spectral density* $\mathscr{A}(\omega)$ of $A(t)$ is defined as the Fourier transform of its autocorrelation function, so that

$$\mathscr{A}(\omega) = \int_{-\infty}^{\infty} \langle A^*(0)A(t)\rangle\, e^{-i\omega t}\, dt. \tag{2.20}$$

Let us put

$$a(\omega) = \int_{0}^{\infty} \langle A^*(0)A(t)\rangle\, e^{-i\omega t}\, dt \tag{2.21}$$

and assume that the autocorrelation function is real. Then from (2.19) and (2.21)

$$a^*(\omega) = \int_{0}^{\infty} \langle A^*(0)A(-t)\rangle\, e^{i\omega t}\, dt$$

$$= \int_{-\infty}^{0} \langle A^*(0)A(t)\rangle\, e^{-i\omega t}\, dt.$$

It follows from (2.20) that

$$\mathscr{A}(\omega) = a(\omega) + a^*(\omega), \tag{2.22}$$

which shows that $\mathscr{A}(\omega)$ is a real function of ω. We deduce from (2.21) and (2.22) that

$$\mathscr{A}(\omega) = 2\int_{0}^{\infty} \langle A^*(0)A(t)\rangle\, \cos \omega t\, dt. \tag{2.23}$$

Equation (2.23) shows that $\mathscr{A}(\omega)$ is also an even function of ω.

The *correlation time* τ_A for $A(t)$ is defined by

$$\tau_A = \int_{0}^{\infty} \frac{\langle A^*(0)A(t)\rangle}{\langle A^*(0)A(0)\rangle}\, dt, \tag{2.24}$$

the integral with respect to t from 0 to ∞ of the normalized autocorrelation function (2.18) of $A(t)$.

When t becomes very great, the value of $B(t'+t)$ in $\langle A^*(t')B(t'+t)\rangle$ becomes independent of the value of $A^*(t')$ on account of the randomness of the motion. We therefore have from (2.10)

$$\lim_{t\to\infty} \langle A^*(t')B(t'+t)\rangle = \lim_{t\to\infty} [\langle A^*(t')\rangle\langle B(t'+t)\rangle],$$

that is,

$$\lim_{t\to\infty} \langle A^*(t')B(t'+t)\rangle = \langle A^*(t')\rangle \lim_{t\to\infty} \langle B(t'+t)\rangle. \tag{2.25}$$

It is therefore clear that $\lim\limits_{t\to\infty} \langle A^*(t')B(t'+t)\rangle$ vanishes, if either $\langle A(\infty)\rangle$ or $\langle B(\infty)\rangle$ vanishes.

2.4 Models of random motion

In this section we explain how probability concepts are applied to random motion. Various models of random motion have been proposed but we shall confine our attention to those based on random flights, on Brownian motion and on Gordon's m- and J-diffusion theories. Our exposition will have nuclear magnetic relaxation in mind but most of the results of this section have wider application. We begin with a discussion of random flights – also called random walk.

2.4.1 Random flights

A theory of random walk applicable to hard identical spherical molecules, each having a spin at its centre, is due to Torrey (1953), who assumed that the positions of the spins are statistically equivalent. Hence the probability of finding a spin, which was initially at the origin, in the element of volume $d^3\mathbf{r}$ at the point $\mathbf{r}$ after one flight is expressible as $P_1(\mathbf{r})d^3\mathbf{r}$. Thus $P_1(\mathbf{r})$ is a probability density function. Similarly the probability of finding it there after n flights may be written $P_n(\mathbf{r})d^3\mathbf{r}$. Then it is known (Chandrasekhar, 1943) that

$$P_n(\mathbf{r}) = \frac{1}{8\pi^3} \int (A(\rho))^n \, e^{i(\mathbf{r}\cdot\rho)} \, d^3\rho, \tag{2.26}$$

where

$$A(\rho) = \int P_1(\mathbf{r}) \, e^{i(\mathbf{r}\cdot\rho)} \, d^3\mathbf{r}. \tag{2.27}$$

On employing the result for the three-dimensional Dirac delta function

$$\delta(\mathbf{r}) = \frac{1}{8\pi^3} \int e^{-i(\mathbf{r}\cdot\rho)} \, d^3\rho \tag{2.28}$$

we see from (2.26) that

$$P_0(\mathbf{r}) = \delta(\mathbf{r}). \tag{2.29}$$

This was to be expected, since the probability of the spin not being at $\mathbf{r}$ is zero and the integral of $\delta(\mathbf{r})$ over all space is unity.

Let us denote by $\mathbf{r}(t)$ the position at time t of a spin in which we are interested, and by $w(\mathbf{r}(0), 0; \mathbf{r}(t), \mathbf{r})$ the conditional probability density that, if the spin is at the origin at time zero, it will be at $\mathbf{r}$ at time t. If $v_n(t)$ is the probability that n flights take place in time t, we have the relation

$$w(\mathbf{r}(0), 0; \mathbf{r}(t), \mathbf{r}) = \sum_{n=0}^{\infty} P_n(\mathbf{r}) v_n(t). \tag{2.30}$$

We assume for $v_n(t)$ the Poisson distribution introduced in (2.3), so that

$$v_n(t) = \left(\frac{t}{\tau}\right)^n \frac{e^{-t/\tau}}{n!}, \tag{2.31}$$

where τ is the mean time between flights. From (2.26) and (2.29)–(2.31) we deduce that

$$w(\mathbf{r}(0), 0; \mathbf{r}(t), \mathbf{r}) = \frac{1}{8\pi^3} \int \exp\left\{ -i(\mathbf{r} \cdot \boldsymbol{\rho}) - \frac{t}{\tau}[1 - A(\rho)] \right\} d^3\rho. \tag{2.32}$$

We next consider two spins i, j with position vectors $\mathbf{r}_i(u)$, $\mathbf{r}_j(u)$ at time u. We write $\mathbf{r}_i, \mathbf{r}_j$ for the position vectors at time t,

$$\mathbf{r} = \mathbf{r}_j - \mathbf{r}_i, \qquad \mathbf{r}' = \tfrac{1}{2}(\mathbf{r}_j + \mathbf{r}_i) \tag{2.33}$$

and $w(\mathbf{r}_j(0) - \mathbf{r}_i(0), \mathbf{R}; \mathbf{r}_j(t) - \mathbf{r}_i(t), \mathbf{r})$ for the conditional probability density that, if at time zero spin j is located at $\mathbf{R}$ relative to spin i, it will at a later time t be located at $\mathbf{r}$ relative to the new position of spin i. Now $w(\mathbf{r}_i(0), 0; \mathbf{r}_i(t), \mathbf{r}_i)$ ensures that, if at time zero spin i is at the origin, it will be at $\mathbf{r}_i$ at time t. Similarly $w(\mathbf{r}_j(0) - \mathbf{R}, 0; \mathbf{r}_j(t) - \mathbf{R}, \mathbf{r}_j - \mathbf{R})$ ensures that, if at time zero spin j is at $\mathbf{R}$, it will be at $\mathbf{r}_j$ at time t. We assume that the two molecules bearing the spins i and j move independently, thus ignoring the possibility that the two molecules may be at the same position at time t. Then the last two conditional probability densities will be independent and so will be multiplied together for the calculation of joint probabilities. Let us now consider

$$w(\mathbf{r}_i(0), 0; \mathbf{r}_i(t), \mathbf{r}_i) w(\mathbf{r}_j(0) - \mathbf{R}, 0; \mathbf{r}_j(t) - \mathbf{R}, \mathbf{r}_j - \mathbf{R}) \, d^3\mathbf{r}_i \, d^3\mathbf{r}_j. \tag{2.34}$$

From (2.33) we deduce that the Jacobian of the transformation from $(\mathbf{r}_i, \mathbf{r}_j)$ to $(\mathbf{r}', \mathbf{r})$ is unity, so that for integration purposes we may replace $d^3\mathbf{r}_i \, d^3\mathbf{r}_j$ in (2.34) by $d^3\mathbf{r}' \, d^3\mathbf{r}$. On doing this we see that

$$w(\mathbf{r}_j(0) - \mathbf{r}_i(0), \mathbf{R}; \mathbf{r}_j(t) - \mathbf{r}_i(t), \mathbf{r}) \, d^3\mathbf{r}$$

$$= \int w(\mathbf{r}_i(0), 0; \mathbf{r}_i(t), \mathbf{r}_i) w(\mathbf{r}_j(0) - \mathbf{R}, 0; \mathbf{r}_j(t) - \mathbf{R}, \mathbf{r}_j - \mathbf{R}) \, d^3\mathbf{r}' \cdot d^3\mathbf{r}. \tag{2.35}$$

To perform the integration over the centre of mass $\mathbf{r}'$ of the two spins we express the two w's in the integrand by (2.32). Then employing (2.28) and (2.33) we deduce from (2.35) that

$$w(\mathbf{r}_j(0) - \mathbf{r}_i(0), \mathbf{R}; \mathbf{r}_j(t) - \mathbf{r}_i(t), \mathbf{r}) = \frac{1}{8\pi^3} \int\int d^3\rho \, d^3\rho' \delta(\rho + \rho')$$

$$\times \exp[-\tfrac{1}{2}i(\mathbf{r} \cdot \rho' - \rho)] \exp[i(\mathbf{R} \cdot \rho')] \exp\left[-\frac{t}{\tau}(2 - A(\rho) - A(\rho'))\right],$$

so that

$$w(\mathbf{r}_j(0) - \mathbf{r}_i(0), \mathbf{R}; \mathbf{r}_j(t) - \mathbf{r}_i(t), \mathbf{r})$$

$$= \frac{1}{8\pi^3} \int d^3\rho \exp[i(\mathbf{r} - \mathbf{R} \cdot \rho)] \exp\left[-\frac{t}{\tau}(2 - A(\rho) - A(-\rho))\right].$$

On putting $\rho \mapsto -\rho$ and noting that $\int d^3(-\rho) = \int d^3\rho$ we express the preceding equation as

$$w(\mathbf{r}_j(0) - \mathbf{r}_i(0), \mathbf{R}; \mathbf{r}_j(t) - \mathbf{r}_i(t), \mathbf{r})$$

$$= \frac{1}{8\pi^3} \int d^3\rho \exp[-i(\mathbf{r} - \mathbf{R}, \rho)] \exp\left[-\frac{t}{\tau}(2 - A(\rho) - A(-\rho))\right]. \quad (2.36)$$

In specific models (Hertz, 1967, p. 177) $A(\rho)$ is often an even function of ρ. When this is so, (2.36) becomes

$$w(\mathbf{r}_j(0) - \mathbf{r}_i(0), \mathbf{R}; \mathbf{r}_j(t) - \mathbf{r}_i(t), \mathbf{r})$$

$$= \frac{1}{8\pi^3} \int d^3\rho \exp[-i(\mathbf{r} - \mathbf{R} \cdot \rho)] \exp\left[-\frac{2t}{\tau}(1 - A(\rho))\right], \quad (2.37)$$

which agrees with Torrey's equation (15). On comparing (2.37) with (2.32) we see that

$$w(\mathbf{r}_j(0) - \mathbf{r}_i(0), \mathbf{R}; \mathbf{r}_j(t) - \mathbf{r}_i(t), \mathbf{r}) = w(\mathbf{r}(0) - \mathbf{R}(0), 0; \mathbf{r}(2t) - \mathbf{R}(2t), \mathbf{r} - \mathbf{R}). \quad (2.38)$$

As an illustration of this theory let us suppose that $P_1(\mathbf{r})$ is a function $P_1(r)$ of r alone. Then, from (2.27), $A(\rho)$ is a function $A(\rho)$ of ρ alone, and expanding the exponential in the integrand we find that

$$A(\rho) = \sum_{n=0}^{\infty} \frac{(i\rho)^n}{n!} \int\int\int P_1(r)r^{n+2} \, dr \cos^n \theta \sin \theta \, d\theta \, d\phi.$$

The integral vanishes for odd values of n and putting $n = 2m$ we obtain

$$A(\rho) = \sum_{m=0}^{\infty} \frac{(-)^m \langle r^{2m} \rangle \rho^{2m}}{(2m+1)!}, \quad (2.39)$$

where

$$\langle r^k \rangle = 4\pi \int_0^{\infty} P_1(r)r^{k+2} \, dr. \quad (2.40)$$

Since the volume element is $4\pi r^2\, dr$, the right hand side gives the mean value of r^k for the probability distribution function $P_1(r)$. We employ in (2.40) the notation for this mean value usually found in the literature, which is allowable to us in the present context since we are not concerned here with ensemble averages. The quantity $\langle r^k\rangle$ is called the *kth moment* of the probability distribution specified by $P_1(r)$; that is to say, it is the mean value of r^k at the end of one flight. Thus $\langle r^2\rangle$ is the *mean squared flight distance*.

Returning to (2.39) we see that $A(\rho)$ is an even function of ρ and on expanding the series we obtain

$$A(\rho)=1-\tfrac{1}{6}\langle r^2\rangle\rho^2+\mathrm{O}(\rho^4).\tag{2.41}$$

Let us assume that in (2.32) the mean squared flight distance is sufficiently small and that t/τ is sufficiently great that, from (2.41), we may approximate $(t/\tau)[1-A(\rho)]$ by $(t/6\tau)\langle r^2\rangle\rho^2$ and write

$$w(\mathbf{r}(0),0;\mathbf{r}(t),\mathbf{r})=\frac{1}{8\pi^3}\int_{-\infty}^{\infty}\exp\left\{-i(\mathbf{r}\cdot\boldsymbol{\rho})-\frac{1}{6}\frac{t}{\tau}\langle r^2\rangle\rho^2\right\}d^3\rho$$

$$=\frac{1}{8\pi^3}\prod_{x,y,z}\int_{-\infty}^{\infty}\cos(x\rho_x)\exp\left[-\frac{1}{6}\frac{t}{\tau}\langle r^2\rangle\rho_x^2\right]d\rho_x.$$

On employing the relation (Peirce, 1929, p. 64)

$$\int_{-\infty}^{\infty}\exp(-a^2x^2)\cos bx\,dx=\frac{\pi^{1/2}\exp\{-b^2/(4a^2)\}}{a}\qquad(a>0)$$

we obtain

$$w(\mathbf{r}(0),0;\mathbf{r}(t),\mathbf{r})=\frac{\exp\left\{-r^2\Big/\left(\dfrac{2t}{3\tau}\langle r^2\rangle\right)\right\}}{8\pi^{3/2}\left(\dfrac{1}{6}\dfrac{t}{\tau}\langle r^2\rangle\right)^{3/2}}.$$

We write this

$$w(\mathbf{r}(0),0;\mathbf{r}(t),\mathbf{r})=\frac{\exp\{-r^2/(4D't)\}}{(4\pi D't)^{3/2}},\tag{2.42}$$

where

$$D'=\frac{\langle r^2\rangle}{6\tau}.\tag{2.43}$$

On calculating $\nabla^2 w$ from

$$\nabla^2 w(\mathbf{r}(0),0;\mathbf{r}(t),\mathbf{r})=\left(\frac{\partial^2}{\partial r^2}+\frac{2}{r}\frac{\partial}{\partial r}\right)w(\mathbf{r}(0),0;\mathbf{r}(t),\mathbf{r})$$

we find that

$$\frac{\partial w(\mathbf{r}(0), 0; \mathbf{r}(t), \mathbf{r})}{\partial t} = D' \nabla^2 w(\mathbf{r}(0), 0; \mathbf{r}(t), \mathbf{r}) \tag{2.44}$$

with the *self-diffusion coefficient D'* defined by (2.43). Equation (2.44) is an example of a *diffusion equation* similar to those found in the conduction of heat or of electricity. It was obtained under the two conditions that the mean squared flight distance is small and that the time t is long compared with the mean time between flights. When the conditional probability density for the motion of system of particles satisfies (2.44), we say that the particles are undergoing *diffusive motion*. Finally on combining (2.38) and (2.42) we obtain

$$w(\mathbf{r}_j(0) - \mathbf{r}_i(0), \mathbf{R}; \mathbf{r}_j(t) - \mathbf{r}_i(t), \mathbf{r}) = \frac{\exp(-|\mathbf{r} - \mathbf{R}|^2 / 8D't)}{(8\pi D't)^{3/2}}. \tag{2.45}$$

This result, like (2.38), is based on the assumption that spin i and spin j move independently of each other.

2.4.2 Brownian motion

A liquid molecule undergoes a constant bombardment from other molecules that produces a rapidly changing motion, both translational and rotational. Translational Brownian motion may be regarded as a limiting case of random walk where $\langle r^2 \rangle$ is extremely small. The theory of Brownian motion is not confined to large values of time. One-dimensional Brownian motion of a particle of mass m may be described by the *Langevin equation* (McConnell, 1980b, Chap. 5)

$$m \frac{du(t)}{dt} = -mB'u(t) + A'(t), \tag{2.46}$$

where $u(t)$ is the velocity, $A'(t)$ is a random driving force independent of $u(t)$ which varies very rapidly, and $mB'u(t)$ is a frictional resisting force. On solving (2.46) it is found that $u(t)$ is a Gaussian random variable; that is to say, $u(t)$ has a probability density function that can be put into the form of $p(X; x)$ in (2.2). Moreover for steady state motion the ensemble average $\langle u(t) \rangle$ of $u(t)$ vanishes and the correlation function

$$\langle u(t)u(s) \rangle = \frac{kT}{m} e^{-B'|t-s|}. \tag{2.47}$$

For the three-dimensional motion of the particle (2.46) is generalized to

$$m\frac{d\mathbf{u}(t)}{dt} = -mB'\mathbf{u}(t) + \mathbf{A}'(t). \tag{2.48}$$

For steady state motion it is found that $\langle \mathbf{u}(t) \rangle$ vanishes and that (2.47) is generalized to

$$\langle u_i(t)u_j(s) \rangle = \delta_{ij}\frac{kT}{m}\,e^{-B'|t-s|}, \tag{2.49}$$

where δ_{ij} is the Kronecker delta whose value is unity for $i=j$ and zero for $i \neq j$. On the other hand, if at time zero the particle is at rest at the origin it may be shown (Chandrasekhar, 1943, p. 25) that the conditional probability density

$$w(\mathbf{r}(0), 0; \mathbf{r}(t), \mathbf{r}) = \left(\frac{mB'^2}{2\pi kT(2B't-3+4\,e^{-B't}-e^{-2B't})}\right)^{3/2}$$

$$\times \exp\left(-\frac{mB'^2r^2}{2kT(2B't-3+4\,e^{-B't}-e^{-2B't})}\right). \tag{2.50}$$

If we take t sufficiently long that $B't \gg 1$, this equation reduces to

$$w(\mathbf{r}(0), 0; \mathbf{r}(t), \mathbf{r}) = \left(\frac{mB'}{4\pi kTt}\right)^{3/2} \exp\left(-\frac{mB'r^2}{4kTt}\right). \tag{2.51}$$

We write this

$$w(\mathbf{r}(0), 0; \mathbf{r}(t), \mathbf{r}) = \frac{\exp\left(-\dfrac{r^2}{4D't}\right)}{(4\pi D't)^{3/2}}, \tag{2.52}$$

where now the self-diffusion coefficient D' is defined by

$$D' = \frac{kT}{mB'}. \tag{2.53}$$

We see that w satisfies

$$\frac{\partial w(r(0), 0; \mathbf{r}(t), \mathbf{r})}{\partial t} = D'\nabla^2 w(\mathbf{r}(0), 0; \mathbf{r}(t), \mathbf{r}). \tag{2.54}$$

This is *Smoluchowski's equation* (von Smoluchowski, 1915) when no external field is present. It is identical with (2.44), if D' is given by (2.53).

Turning to rotational Brownian motion we consider a spherical molecule with moment of inertia I about a diameter, which is being tossed around in a heat bath by a random couple of moment $\mathbf{A}(t)$ that is independent of the angular velocity $\omega(t)$ of the molecule. The Langevin equation corresponding to (2.48) is

$$I\frac{d\omega(t)}{dt} = -IB\omega(t) + \mathbf{A}(t), \tag{2.55}$$

where IB is the coefficient of friction for the resisting couple. The angular velocity components are Gaussian random variables. For steady state motion the ensemble average $\langle \omega(t) \rangle$ vanishes and

$$\langle \omega_i(t)\omega_j(s) \rangle = \delta_{ij} \frac{kT}{I} e^{-B|t-s|}. \tag{2.56}$$

We deduce from (2.24) and (2.56) that the correlation time for $\omega_i(t)$ is B^{-1}. This is often called the *friction time* and denoted by τ_F.

All our calculations have so far been performed in a laboratory coordinate system which we shall call S. In order to study the rotational Brownian motion of an asymmetric rigid molecule we take a body coordinate system S' with origin at the centre of mass and axes in the directions of the principal axes of inertia of the molecule. We then generalize (2.55) to the *Euler–Langevin equations*:

$$I_1 \frac{d\omega_1}{dt} - (I_2 - I_3)\omega_2\omega_3 = -I_1 B_1 \omega_1 + A_1(t),$$

$$I_2 \frac{d\omega_2}{dt} - (I_3 - I_1)\omega_3\omega_1 = -I_2 B_2 \omega_2 + A_2(t), \tag{2.57}$$

$$I_3 \frac{d\omega_3}{dt} - (I_1 - I_2)\omega_1\omega_2 = -I_3 B_3 \omega_3 + A_3(t).$$

In these equations I_1, I_2, I_3 are the principal moments of inertia, $\omega_1, \omega_2, \omega_3$ are the respective components of angular velocity, and B_1, B_2, B_3 are the respective friction constants. The left hand sides of (2.57) are expressions for the rate of change of the components of angular momentum about the rotating coordinate axes. On the right hand sides we have the components $A_1(t)$, $A_2(t)$, $A_3(t)$ of the thermal driving couple and the components $-I_1 B_1 \omega_1$, $-I_2 B_2 \omega_2$, $-I_3 B_3 \omega_3$ of the frictional couple, which are assumed to be proportional to the respective angular velocity components. The correlation function of $\omega_i(t)$ and $\omega_j(s)$ vanishes for $i \neq j$, and for $i = j$ it is expressible as a series of which the first term has the form of the right hand side of (2.56) and the remaining terms provide a small correction. When evaluating *orientational correlation functions*, that is those for which the $A(s(t))$ and $B(s(t))$ of section 2.3 depend on angular variables, it has been found advantageous to use the value of the angular velocity correlation function $\langle \omega_i(t)\omega_j(s) \rangle$ in order to first calculate the operator $R(t)$ which corresponds to the rotation of the body coordinate system S' from its orientation at time zero to its orientation at time t (Ford *et al.*, 1979). This is explained more fully in Appendixes C and D.

The analytical discussion of the consequences of the Euler–Langevin equations for rotational Brownian motion can be very tedious. The same is true for the consequences of (2.50) for translational Brownian motion. Hence (2.51) is often employed in calculations, and in fact the ensuing results are frequently quite valuable. We deduced (2.51) from (2.50) as a long time limit. We can also deduce it by applying the limiting process

$$m \to 0, \qquad B' \to \infty, \qquad mB' \text{ finite}. \tag{2.58}$$

On comparing (2.55) with (2.48) we see that the analogous limiting procedure for the rotational motion of a sphere is

$$I \to 0, \qquad B \to \infty, \qquad IB \text{ finite}. \tag{2.59}$$

This describes what Debye did in order to study the rotational Brownian motion of a polar spherical molecule. Indeed for uniplanar rotation his equation would read in our notation

$$I\ddot{\theta} = -\mu F \sin\theta - \zeta\dot{\theta} + A(t) \qquad (\zeta = IB), \tag{2.60}$$

where F is an applied external electric field, μ is the electric dipole moment and θ is the angle between the direction of F and the dipole axis. To solve this equation Debye (1929, p. 82) neglected the $I\ddot{\theta}$-term. Since the friction constant ζ is finite, Debye's method is to introduce the limiting processes of (2.59) into (2.60). We therefore define the *Debye approximation* or *Debye limit* of any result as that which would be obtained for translational motion from (2.58) and for rotational motion from (McConnell, 1984b)

$$I_i \to 0, \qquad B_i \to \infty, \qquad I_i B_i \text{ finite} \qquad (i = 1, 2, 3). \tag{2.61}$$

We shall designate a theory in which such limits are not introduced as an *inertial theory* or as a theory including *inertial effects*. A calculation in the Debye approximation is thus one that ignores inertial effects. A theory in which inertial effects are ignored is also called a *diffusion theory* and, if it deals with rotational motion, it is a *rotational diffusion theory*.

2.4.3 Extended diffusion

The basic idea of rotational Brownian motion is that the particle in question undergoes a very frequent succession of infinitesimal angular displacement. Gordon (1966) pointed out that this description may not be valid for small molecules in solution, and he considered the possibility of a molecule rotating freely through a finite angle between collisions. He then proposed two models:

(a) *m-diffusion*. In this model the components of the angular momentum **J** of molecular rotation are randomized at the end of a free rotational step. However, the magnitude of **J** and the orientation of the molecule are unaltered by the collision.

(*b*) *J*-diffusion. This model differs from the previous one in that at the end of a free rotational step the magnitude of **J** is distributed according to the Boltzmann law.

Gordon takes a unit vector **U** fixed in the molecule and gives a general expression for its correlation function $\langle(\mathbf{U}(0)\cdot\mathbf{U}(t))\rangle$ in the two models. The calculations cannot be completed by analytical methods.

2.4.4 *General assumptions in future calculations*

When performing calculations on nuclear magnetic relaxation we shall make the following assumptions:

(*a*) The influence of internal rotations and of vibrations in molecules may be neglected. This justifies the use of the dynamics of a rigid body in writing down (2.57).

(*b*) Translational and rotational motion may be investigated independently of one another; in other words, we neglect rototranslational effects.

A well known theorem in the kinematics of a rigid body states that the rotation of the body about any axis is equivalent to an equal rotation about a parallel axis together with a translation. This theorem is easily established by putting two equal and opposite rotations about the second axis. Under assumption (*b*) we deduce that, if we change the origin of the body coordinate system from the centre of mass to any other point fixed in the molecule, the operator $R(t)$ of subsection 2.4.2 will still describe the rotation of the body coordinate system at time zero to its orientation at time t. This result will be used in Chapters 6, 8 and 9, when we shall discuss nuclear magnetic relaxation by intramolecular interactions.

3

Equations of relaxation theory

3.1 The density operator

The present chapter will be devoted to the discussion of topics in quantum theory, which will be required for application to the various relaxation processes that will be investigated in later chapters. Frequent use will be made of the standard formalism of quantum mechanics. A summary of this is to be found in Appendix A, and results from there will be employed without explicit reference.

Probability density functions have already been introduced in Chapter 2. When we have a quantum mechanical system in a state described by the normalized wave function ψ, the probability density is given by Born's expression $\psi^* \psi$. Let us expand ψ in terms of a time independent basis $f_1, f_2, f_3, \ldots$ as

$$\psi = c_1 f_1 + c_2 f_2 + c_3 f_3 + \cdots \tag{3.1}$$

where $c_1, c_2, c_3, \ldots$ may depend on the time. The *expectation value* of a physical observable described by the self-adjoint operator A is $\int \psi^* A \psi \, dq$, where q denotes the real continuous variables of which ψ is a function. From (3.1)

$$\int \psi^* A \psi \, dq = \sum_{m,n} c_m^* c_n \int f_m^* A f_n \, dq,$$

so that

$$\int \psi^* A \psi \, dq = \sum_{n,m} c_n c_m^* A_{mn}, \tag{3.2}$$

where

$$A_{mn} = \int f_m^* A f_n \, dq. \tag{3.3}$$

Let us denote by σ the operator such that

$$\sigma_{nm} = \int f_n^* \sigma f_m \, dq = c_n c_m^*. \tag{3.4}$$

Since the matrix composed of the elements σ_{nm} is obviously Hermitian, σ is a self-adjoint operator. Moreover from (3.2) and (3.4)

$$\int \psi^* A \psi \, dq = \sum_{n,m} \sigma_{nm} A_{mn} = \sum_n (\sigma A)_{nn}.$$

This result, which is independent of the chosen basis $f_1, f_2, f_3, \ldots$, may be expressed by

$$\int \psi^* A \psi \, dq = \text{tr}(\sigma A). \tag{3.5}$$

Let us relate these considerations with those in the previous chapter. When we are dealing with random motion, the coefficients $c_1, c_2, c_3, \ldots$ in (3.1) are random variables, whether they are time dependent or not. In order to construct an ensemble average, we take an average over the values of $c_1, c_2, c_3, \ldots$ at one instant of time. Since, from (3.3), the formation of matrix elements is an operation independent of the c's, we deduce from (3.5) that the ensemble average of the expectation value

$$\left\langle \int \psi^* A \psi \, dq \right\rangle = \langle \text{tr}(\sigma A) \rangle = \text{tr}(\langle \sigma \rangle A). \tag{3.6}$$

We define the *density operator* ρ by

$$\rho = \langle \sigma \rangle. \tag{3.7}$$

Hence, from (3.6) and (3.7),

$$\left\langle \int \psi^* A \psi \, dq \right\rangle = \text{tr}(\rho A), \tag{3.8}$$

which shows that the density operator may be very useful for calculating ensemble averages. The matrix representative of ρ for any given basis is the *density matrix* for that basis.

We now consider the time dependence of the density operator. If $h\mathcal{H}$ is the self-adjoint Hamiltonian operator for the system, the equation of matrix mechanics

$$\frac{d\rho(t)}{dt} = \frac{i}{h} [\rho(t), h\mathcal{H}]$$

simplifies to

$$\frac{d\rho(t)}{dt} = i[\rho(t), \mathcal{H}]. \tag{3.9}$$

We shall show that this equation is satisfied for a time independent $\mathcal{H}$ by

$$\rho(t) = e^{-i\mathcal{H}t} \rho(0) e^{i\mathcal{H}t}, \tag{3.10}$$

where the exponential of an operator O is defined by

$$e^O = E + O + \frac{O^2}{2!} + \frac{O^3}{3!} + \cdots \tag{3.11}$$

and E is the identity operator. For the product of three time dependent operators $B(t)$, $C(t)$, $D(t)$

$$\frac{d}{dt}(B(t)C(t)D(t)) = \lim_{\delta t \to 0} \{(B(t+\delta t)C(t+\delta t)D(t+\delta t) - B(t)C(t)D(t))/\delta t\}.$$

Expanding

$$B(t+\delta t) = B(t) + \delta t \frac{dB(t)}{dt} + \cdots$$

etc., we easily find that

$$\frac{d}{dt}(B(t)C(t)D(t)) = \frac{dB(t)}{dt}C(t)D(t) + B(t)\frac{dC(t)}{dt}D(t) + B(t)C(t)\frac{dD(t)}{dt}. \quad (3.12)$$

On applying this to (3.10) and noting from (3.11) that

$$\frac{d}{dt}(e^{\pm i\mathcal{H}t}) = \pm i\mathcal{H} e^{\pm i\mathcal{H}t} = \pm i e^{\pm i\mathcal{H}t}\mathcal{H} \quad (3.13)$$

we find that (3.10) is the solution of (3.9). It is clear that $\rho(t)$ is obtained in (3.10) by a unitary, or canonical, transformation of $\rho(0)$.

Let us construct the matrix representative of $\rho(t)$ with respect to the basis constituted by the eigenfunctions of the time independent operator $\mathcal{H}$. If $E_1, E_2, E_3, \ldots$ are the energy levels, and $f_1, f_2, f_3, \ldots$ the respective basis elements, so that

$$\hbar\mathcal{H}f_1 = E_1 f_1, \qquad \hbar\mathcal{H}f_2 = E_2 f_2, \quad (3.14)$$

etc., the matrix element

$$(m|\rho(t)|n) \equiv \int f_m^* \rho(t) f_n \, dq$$

$$= \int f_m^* e^{-i\mathcal{H}t} \rho(0) e^{i\mathcal{H}t} f_n \, dq$$

$$= \int (e^{i\mathcal{H}t} f_m)^* \rho(0) e^{i\mathcal{H}t} f_n \, dq.$$

Now, from (3.11) and (3.14)

$$e^{i\mathcal{H}t} f_n = \left(E + \frac{itE_n}{\hbar} + \frac{i^2 t^2 E_n^2}{2\hbar^2} + \cdots\right)f_n = \exp(iE_n t/\hbar)f_n$$

and therefore

$$(m|\rho(t)|n) = \exp\left[\frac{i}{\hbar}(E_n - E_m)t\right]\int f_m^* \rho(0) f_n \, dq.$$

Thus we have

$$(m|\rho(t)|n) = \exp\left[\frac{i}{\hbar}(E_n - E_m)t\right](m|\rho(0)|n). \quad (3.15)$$

3.2 Interaction representation

We return to the problem of solving (3.9) for the density operator $\rho(t)$. We no longer assume that the Hamiltonian $\hbar\mathcal{H}$ is time independent but we suppose that it is expressible by

$$\hbar\mathcal{H} = \hbar\mathcal{H}_0 + \hbar G(t), \tag{3.16}$$

where $\hbar\mathcal{H}_0$ is a large time independent interaction Hamiltonian, and $\hbar G(t)$ is a comparatively small time dependent one which we regard as producing a small perturbation. Both $\mathcal{H}_0$ and $G(t)$ are self-adjoint operators. Equations (3.9) and (3.16) yield

$$\frac{d\rho(t)}{dt} = i[\rho(t), \mathcal{H}_0 + G(t)]. \tag{3.17}$$

We define the *interaction representation* $A(t)^{int}$ of an operator $A(t)$ as that given by the canonical transformation

$$A(t)^{int} = e^{i\mathcal{H}_0 t} A(t) e^{-i\mathcal{H}_0 t}, \tag{3.18}$$

so that

$$A(t) = e^{-i\mathcal{H}_0 t} A(t)^{int} e^{i\mathcal{H}_0 t}. \tag{3.19}$$

Thus we have

$$G(t)^{int} = e^{i\mathcal{H}_0 t} G(t) e^{-i\mathcal{H}_0 t} \tag{3.20}$$

$$\rho(t)^{int} = e^{i\mathcal{H}_0 t} \rho(t) e^{-i\mathcal{H}_0 t} \tag{3.21}$$

$$\rho(0)^{int} = \rho(0). \tag{3.22}$$

From (3.12), (3.13) and (3.19) we have

$$\frac{d\rho(t)}{dt} = \frac{d}{dt} (e^{-i\mathcal{H}_0 t} \rho(t)^{int} e^{i\mathcal{H}_0 t})$$

$$= i[\rho(t), \mathcal{H}_0] + e^{-i\mathcal{H}_0 t} \frac{d\rho(t)^{int}}{dt} e^{i\mathcal{H}_0 t}.$$

Comparing this with (3.17) we deduce that

$$e^{-i\mathcal{H}_0 t} \frac{d\rho(t)^{int}}{dt} e^{i\mathcal{H}_0 t} = i[\rho(t), G(t)],$$

so that

$$\frac{d\rho(t)^{int}}{dt} = i e^{i\mathcal{H}_0 t} [\rho(t), G(t)] e^{-i\mathcal{H}_0 t}$$

and, on employing (3.20) and (3.21),

$$\frac{d\rho(t)^{int}}{dt} = i[\rho(t)^{int}, G(t)^{int}]. \tag{3.23}$$

This is an equation with all the operators expressed in interaction representation.

By employing the method used in the derivation of (3.15) we deduce from (3.19) that

$$(m|A(t)|n) = \exp\left[\frac{i}{\hbar}(E_n - E_m)t\right](m|A(t)^{\text{int}}|n), \qquad (3.24)$$

where $|m)$, $|n)$ are eigenfunctions of $\hbar\mathcal{H}_0$ with respective eigenvalues E_m, E_n. From (3.24) we deduce that

$$(m|G(t)|n) = \exp\left[\frac{i}{\hbar}(E_n - E_m)t\right](m|G(t)^{\text{int}}|n) \qquad (3.25)$$

$$(m|\rho(t)|n) = \exp\left[\frac{i}{\hbar}(E_n - E_m)t\right](m|\rho(t)^{\text{int}}|n). \qquad (3.26)$$

The solution of (3.23) may formally be expressed by

$$\rho(t)^{\text{int}} = \rho(0)^{\text{int}} + i \int_0^t \left[\rho(t')^{\text{int}}, G(t')^{\text{int}}\right] dt'. \qquad (3.27)$$

This clearly satisfies (3.23) but otherwise tells us little, since the integrand contains the unknown function. We therefore proceed by successive approximation firstly replacing $\rho(t')^{\text{int}}$ by $\rho(0)^{\text{int}}$, which by (3.22) is $\rho(0)$, and obtaining

$$\rho(t)^{\text{int}} = \rho(0) + i \int_0^t \left[\rho(0), G(t')^{\text{int}}\right] dt',$$

so that

$$\rho(t')^{\text{int}} = \rho(0) + i \int_0^{t'} \left[\rho(0), G(t'')^{\text{int}}\right] dt''. \qquad (3.28)$$

For the next approximation we substitute into (3.27) the value of $\rho(t')^{\text{int}}$ given by (3.28) and deduce that

$$\rho(t)^{\text{int}} = \rho(0) + i \int_0^t \left[\rho(0), G(t')^{\text{int}}\right] dt'$$

$$- \int_0^t dt' \int_0^{t'} dt'' \left[\left[\rho(0), G(t'')^{\text{int}}\right], G(t')^{\text{int}}\right]. \qquad (3.29)$$

In this way we obtain a series of *time ordered integrals*, namely multiple integrals in which the upper limit of integration at each stage is the integration variable for the next stage. The integrands are multiple commutators whose multiplicities increase by unity for each successive approximation. The pattern of the approximations is obvious, but we shall limit our approximations to (3.29).

In our relaxational problems the Hamiltonian $h\mathcal{H}_0$ will arise from a constant field in a fixed direction, and the motion of the molecules will be in a steady state. We may therefore apply to $G(t)^{\text{int}}$ the ergodic theorem, which from (2.6) and (2.7) gives

$$\langle G(t)^{\text{int}}\rangle = \lim_{\tau \to \infty} \frac{1}{\tau} \int_{-\tau/2}^{\tau/2} G(t)^{\text{int}}\, dt.$$

Thus $\langle G(t)^{\text{int}}\rangle$ is a constant and a small one by our earlier supposition. We now add $\langle G(t)^{\text{int}}\rangle$ to $\mathcal{H}_0$ and subtract it from $G(t)^{\text{int}}$. To avoid introducing more variables we shall refer to the new time independent Hamiltonian as $h\mathcal{H}_0$ and to $G(t)^{\text{int}} - \langle G(t)^{\text{int}}\rangle$ as $G(t)^{\text{int}}$, so that we now have

$$\langle G(t)^{\text{int}}\rangle = 0. \tag{3.30}$$

It is evident that, if we had started with a time independent perturbation Hamiltonian hG in (3.16), its net effect would be to produce a constant shift in the energy levels of the system. We shall describe in the next two sections how the time dependence of hG leads to the nuclear magnetic relaxation introduced in section 1.1, and to precise expressions for the longitudinal and transverse relaxation times T_1 and T_2.

3.3 The Redfield equations

We deduce from (3.29) that

$$\frac{d\rho(t)^{\text{int}}}{dt} = i[\rho(0), G(t)^{\text{int}}] - \int_0^t dt'\, [[\rho(0), G(t')^{\text{int}}], G(t)^{\text{int}}], \tag{3.31}$$

where the ensemble average of $G(t)^{\text{int}}$ vanishes by (3.30). We employ the eigenfunctions of the redefined time independent $\mathcal{H}_0$ to construct matrix representatives of operators. Writing $h\omega_l$ for the eigenvalue of $h\mathcal{H}_0$ corresponding to the state with eigenfunction $|l\rangle$ we have from (3.22), (3.25) and (3.26)

$$(m|G(t)^{\text{int}}|n) = \exp[i(\omega_m - \omega_n)t](m|G(t)|n), \tag{3.32}$$

$$(m|\rho(t)^{\text{int}}|n) = \exp[i(\omega_m - \omega_n)t](m|\rho(t)|n), \tag{3.33}$$

$$(m|\rho(0)^{\text{int}}|n) = (m|\rho(0)|n). \tag{3.34}$$

We shall now take ensemble averages of both sides of (3.31). Since $\rho(t)$, and consequently $\rho(t)^{\text{int}}$, are already ensemble averages we omit the angle brackets on the left hand side. Moreover, it follows from (3.30) that

$$\langle [\rho(0)^{\text{int}}, G(t)^{\text{int}}]\rangle = \rho(0)\langle G(t)^{\text{int}}\rangle - \langle G(t)^{\text{int}}\rangle \rho(0) = 0.$$

Then writing $t' = t - \tau$, constructing the $\alpha\alpha'$-element of each side, and employing (3.32)–(3.34) and the multiplication rule for matrices we deduce from (3.31)

$$\frac{d\rho(t)^{int}_{\alpha\alpha'}}{dt} = \sum_{\beta\beta'} \int_0^t d\tau$$

$$\times \langle [G_{\alpha\beta}(t-\tau)\rho(0)_{\beta\beta'}G_{\beta'\alpha'}(t) \exp[-i(\omega_\alpha - \omega_\beta)\tau] \exp[i(\omega_\alpha - \omega_\beta + \omega_{\beta'} - \omega_{\alpha'})t]$$

$$+ G_{\alpha\beta}(t)\rho(0)_{\beta\beta'}G_{\beta'\alpha'}(t-\tau) \exp[i(\omega_{\alpha'} - \omega_{\beta'})\tau] \exp[i(\omega_\alpha - \omega_\beta + \omega_{\beta'} - \omega_{\alpha'})t]$$

$$- \rho(0)_{\alpha\beta}G_{\beta\beta'}(t-\tau)G_{\beta'\alpha'}(t) \exp[i(\omega_{\beta'} - \omega_\beta)\tau] \exp[i(\omega_\beta - \omega_{\alpha'})t]$$

$$- G_{\alpha\beta}(t)G_{\beta\beta'}(t-\tau) \exp[i(\omega_{\beta'} - \omega_\beta)\tau] \exp[i(\omega_\alpha - \omega_{\beta'})t]\rho(0)_{\beta'\alpha'}] \rangle. \quad (3.35)$$

The first term in the square bracket of (3.35) is proportional to $G_{\alpha\beta\alpha'\beta'}(\tau)$, where

$$G_{\alpha\beta\alpha'\beta'}(\tau) = \langle G_{\alpha\beta}(t-\tau)G_{\beta'\alpha'}(t) \rangle. \quad (3.36)$$

Since the motion is steady,

$$G_{\alpha\beta\alpha'\beta'}(\tau) = \langle G_{\alpha\beta}(0)G_{\beta'\alpha'}(\tau) \rangle. \quad (3.37)$$

Since G is self-adjoint, its matrix representative is Hermitian and so

$$G_{\alpha\beta\alpha'\beta'}(\tau) = \langle G^*_{\beta\alpha}(0)G_{\beta'\alpha'}(\tau) \rangle,$$

the cross-correlation function of $G_{\beta\alpha}$ and $G_{\beta'\alpha'}$.

We shall now study the operator G. In doing this we shall refer to properties of spherical tensors that are discussed in Appendixes B and C. For nuclear magnetic relaxation processes it will usually be found that G is expressible by

$$G(t) = \sum_{q=-j}^{j} (-)^q H_{-q}(t) A^{(q)}, \quad (3.38)$$

where $j = 0$, 1 or 2, $A^{(q)}$ is a spin dependent operator and the randomness of $G(t)$ is situated in the spin independent commuting function $H_{-q}(t)$. While G is self-adjoint, the same need not be true for $A^{(q)}$. However, there are certain restrictions on $H_{-q}(t)$ and $A^{(q)}$. From (3.38)

$$G(t) = H_0(t)A^{(0)} - H_{-1}(t)A^{(1)} - H_1(t)A^{(-1)} + H_{-2}(t)A^{(2)} + H_2(t)A^{(-2)}$$

$$G^+(t) = H_0^*(t)A^{(0)+} - H_{-1}^*(t)A^{(1)+} - H_1^*(t)A^{(-1)+} + H_{-2}^*(t)A^{(2)+}$$
$$+ H_2^*(t)A^{(-2)+}$$

and we may have $G^+(t) = G(t)$ by putting

$$A^{(-q)} = A^{(q)+}, \qquad H_{-q}(t) = H_q^*(t)$$

or

$$A^{(-q)} = (-)^q A^{(q)+}, \qquad H_{-q}(t) = (-)^q H_q^*(t). \quad (3.39)$$

To conform with equation (B.3) in Appendix B for spherical harmonics we adopt the convention of (3.39). While the second equation of (3.39) is obeyed by spherical tensor components, the self-adjointness of G does not require that $H_q(t)$ be components of a spherical tensor. It is seen from (B.24), (B.27) and (3.39) that the right hand side of (3.38) is the scalar product of $\mathbf{H}^j(t)$ and $\mathbf{A}^j$.

If $G(t)$ is invariant under three-dimensional rotations and if $A^{(q)}$ are components of a spherical tensor, it will follow from Appendix B that $H_q(t)$ are also components of a spherical tensor. This situation is realized in all the relaxation processes that we shall examine.

On substituting (3.38) into (3.37) we obtain

$$G_{\alpha\beta\alpha'\beta'}(\tau) = \sum_{q,q'=-j}^{j} (-)^{q+q'} \langle H_{-q}(0)H_{-q'}(\tau)\rangle A_{\alpha\beta}^{(q)} A_{\beta'\alpha'}^{(q')}. \tag{3.40}$$

It follows from (C.28) that

$$\langle H_{-q}(0)H_{-q'}(\tau)\rangle = (-)^q \delta_{q,-q'} \langle H_0(0)H_0(\tau)\rangle \tag{3.41}$$

and therefore that

$$G_{\alpha\beta\alpha'\beta'}(\tau) = \langle H_0(0)H_0(\tau)\rangle \sum_{q=-j}^{j} A_{\alpha\beta}^{(q)} A_{\beta'\alpha'}^{(q)+}, \tag{3.42}$$

by (3.39). This last equation also shows that $H_0(t)$ is real. Moreover, since $H_0(t)$ is a commuting quantity and we are dealing with steady state motion,

$$\langle H_0(0)H_0(-\tau)\rangle = \langle H_0(\tau)H_0(0)\rangle = \langle H_0(0)H_0(\tau)\rangle.$$

Hence we conclude from (3.41) that $\langle H_{-q}(0)H_{-q'}(\tau)\rangle$ is a real and even function of τ which vanishes for $q' \neq -q$ and which is independent of q. We then deduce from (3.37) and (3.42) that $G_{\alpha\beta\alpha'\beta'}(\tau)$ is an even function of τ and that

$$G_{\alpha\beta\alpha'\beta'}(\tau) = G_{\alpha\beta\alpha'\beta'}(-\tau) = \langle G_{\alpha\beta}(0)G_{\beta'\alpha'}(-\tau)\rangle$$
$$= \langle G_{\alpha\beta}(\tau)G_{\beta'\alpha'}(0)\rangle = \langle G_{\beta'\alpha'}(0)G_{\alpha\beta}(\tau)\rangle,$$

since matrix elements are commuting quantities. Hence

$$G_{\alpha\beta\alpha'\beta'}(\tau) = G_{\beta'\alpha'\beta\alpha}(\tau), \tag{3.43}$$

and we then see that (3.42) yields the relation

$$\sum_{q=-j}^{j} A_{\beta'\alpha'}^{(q)} A_{\alpha\beta}^{(q)+} = \sum_{q=-j}^{j} A_{\alpha\beta}^{(q)} A_{\beta'\alpha'}^{(q)+}. \tag{3.44}$$

For intramolecular interactions depending on random rotational motion we define the *correlation time* for the relaxation process described by the Hamiltonian $hG(t)$ of (3.38) as the correlation time for $H_q(t)$, as calculated by (2.24).

We return to the evaluation of the integral of (3.35). Since $\rho(0)$ is itself an ensemble average,

$$\langle G_{\alpha\beta}(t-\tau)\rho(0)_{\beta\beta'}G_{\beta'\alpha'}(t)\rangle = \rho(0)_{\beta\beta'}G_{\alpha\beta\alpha'\beta'}(\tau).$$

Let us therefore consider the integral

$$\int_0^t G_{\alpha\beta\alpha'\beta'}(\tau)\exp[-i(\omega_\alpha-\omega_\beta)\tau]\,d\tau. \tag{3.45}$$

On account of the randomness of the motion $G(t-\tau)$ will be independent of $G(t)$ and consequently

$$G_{\alpha\beta\alpha'\beta'}(\tau)=\langle G_{\alpha\beta}(t-\tau)\rangle\langle G_{\beta'\alpha'}(t)\rangle=0$$

by (3.30), unless τ is smaller than a fixed quantity τ_c. We take $\tau<\tau_c$ and we assume that $t\gg\tau_c$. Then

$$\int_t^\infty G_{\alpha\beta\alpha'\beta'}(\tau)\exp[-i(\omega_\alpha-\omega_\beta)\tau]\,d\tau$$

is negligible and we may replace (3.45) by

$$\int_0^\infty G_{\alpha\beta\alpha'\beta'}(\tau)\exp[-i(\omega_\alpha-\omega_\beta)\tau]\,d\tau. \tag{3.46}$$

Let us consider $J_{\alpha\beta\alpha'\beta'}(\omega)$ defined by

$$J_{\alpha\beta\alpha'\beta'}(\omega)=\frac{1}{2}\int_{-\infty}^\infty G_{\alpha\beta\alpha'\beta'}(\tau)\,e^{-i\omega\tau}\,d\tau. \tag{3.47}$$

Since $G_{\alpha\beta\alpha'\beta'}(\tau)$ is an even function,

$$J_{\alpha\beta\alpha'\beta'}(\omega)=\int_0^\infty G_{\alpha\beta\alpha'\beta'}(\tau)\cos\omega\tau\,d\tau$$

and this shows that $J_{\alpha\beta\alpha'\beta'}(\omega)$ is an even function of ω. To discuss it we introduce $j(\omega)$, the spectral density of $2^{-1/2}H_0(t)$ defined according to (2.20) by

$$j(\omega)=\frac{1}{2}\int_{-\infty}^\infty \langle H_0(0)H_0(\tau)\rangle\,e^{-i\omega\tau}\,d\tau, \tag{3.48}$$

since $H_0(0)$ is real by (3.39). The function $j(\omega)$ is clearly an even function of ω and therefore

$$j(\omega)=\mathrm{Re}\left\{\int_0^\infty \langle H_0(0)H_0(\tau)\rangle\,e^{-i\omega\tau}\,d\tau\right\}.$$

If we write

$$\mathrm{Im}\left\{\int_0^\infty \langle H_0(0)H_0(\tau)\rangle\,e^{-i\omega\tau}\,d\tau\right\}$$

$$=-\int_0^\infty \langle H_0(0)H_0(\tau)\rangle\sin\omega\tau\,d\tau\equiv j'(\omega),$$

an odd function of ω, then

$$\int_0^\infty \langle H_0(0)H_0(\tau)\rangle\,e^{-i\omega\tau}\,d\tau=j(\omega)+ij'(\omega).$$

From (3.42), (3.47) and (3.48)

$$J_{\alpha\beta\alpha'\beta'}(\omega)=j(\omega)\sum_{q=-j}^{j}A_{\alpha\beta}^{(q)}A_{\beta'\alpha'}^{(q)+}. \tag{3.49}$$

Moreover

$$\int_{0}^{\infty}G_{\alpha\beta\alpha'\beta'}(\tau)\,\mathrm{e}^{-\mathrm{i}\omega\tau}\,\mathrm{d}\tau=(j(\omega)+\mathrm{i}j'(\omega))\sum_{q=-j}^{j}A_{\alpha\beta}^{(q)}A_{\beta'\alpha'}^{(q)+} \tag{3.50}$$

and we conclude from (3.45) and (3.46) that the first integral on the right hand side of (3.35) is equal to

$$\rho(0)_{\beta\beta'}(j(\omega_{\alpha}-\omega_{\beta})+\mathrm{i}j'(\omega_{\alpha}-\omega_{\beta}))\sum_{q=-j}^{j}A_{\alpha\beta}^{(q)}A_{\beta'\alpha'}^{(q)+}\,\exp[\mathrm{i}(\omega_{\alpha}-\omega_{\beta}+\omega_{\beta'}-\omega_{\alpha'})t]. \tag{3.51}$$

Since $G_{\alpha\beta\alpha'\beta'}(\tau)$ is an even function, we deduce that the second integral of (3.35) is equal to

$$\rho(0)_{\beta\beta'}(j(\omega_{\alpha'}-\omega_{\beta'})+\mathrm{i}j'(\omega_{\alpha'}-\omega_{\beta'}))\sum_{q=-j}^{j}A_{\beta'\alpha'}^{(q)}A_{\alpha\beta}^{(q)+}\,\exp[\mathrm{i}(\omega_{\alpha}-\omega_{\beta}+\omega_{\beta'}-\omega_{\alpha'})t],$$

which by (3.44) is also equal to

$$\rho(0)_{\beta\beta'}(j(\omega_{\alpha'}-\omega_{\beta'})+\mathrm{i}j'(\omega_{\alpha'}-\omega_{\beta'}))$$
$$\times\sum_{q=-j}^{j}A_{\alpha\beta}^{(q)}A_{\beta'\alpha'}^{(q)+}\,\exp[\mathrm{i}(\omega_{\alpha}-\omega_{\beta}+\omega_{\beta'}-\omega_{\alpha'})t]. \tag{3.52}$$

The third integral on the right hand side of (3.35) is equal to

$$-\sum_{\beta\beta'}\rho(0)_{\alpha\beta}\int_{0}^{\infty}G_{\beta\beta'\alpha'\beta'}(\tau)\exp[-\mathrm{i}(\omega_{\beta}-\omega_{\beta'})\tau]\,\mathrm{d}\tau\,\exp[\mathrm{i}(\omega_{\beta}-\omega_{\alpha'})t],$$

that is,

$$-\sum_{\beta\beta'}\rho(0)_{\alpha\beta}(j(\omega_{\beta}-\omega_{\beta'})+\mathrm{i}j'(\omega_{\beta}-\omega_{\beta'}))\sum_{q=-j}^{j}A_{\beta\beta'}^{(q)}A_{\beta'\alpha'}^{(q)+}\,\exp[\mathrm{i}(\omega_{\beta}-\omega_{\alpha'})t]. \tag{3.53}$$

We replace the summation indices β,β' by σ,β' by making the changes

$$\beta'\mapsto\sigma,\qquad\beta\mapsto\beta'$$

and expressing (3.53) as

$$-\sum_{\beta'\sigma}\rho(0)_{\alpha\beta'}(j(\omega_{\beta'}-\omega_{\sigma})+\mathrm{i}j'(\omega_{\beta'}-\omega_{\sigma}))\sum_{q=-j}^{j}A_{\beta'\sigma}^{(q)}A_{\sigma\alpha'}^{(q)+}\,\exp[\mathrm{i}(\omega_{\beta'}-\omega_{\alpha'})t].$$

This is equivalent to

$$-\sum_{\beta\beta'\sigma}\delta_{\alpha\beta}\,\rho(0)_{\beta\beta'}\sum_{\sigma}(j(\omega_{\sigma}-\omega_{\beta'})-\mathrm{i}j'(\omega_{\sigma}-\omega_{\beta'}))$$
$$\times\sum_{q=-j}^{j}A_{\sigma\alpha'}^{(q)}A_{\beta'\sigma}^{(q)+}\,\exp[\mathrm{i}(\omega_{\alpha}-\omega_{\beta}+\omega_{\beta'}-\omega_{\alpha'})t], \tag{3.54}$$

where we have used (3.44) and the relation $\omega_\alpha - \omega_\beta = 0$ which follows from the presence of the Kronecker delta. Similarly we find that the fourth integral on the right hand side of (3.35) is equal to

$$-\sum_{\beta\beta'\sigma} \delta_{\alpha'\beta'} \sum_\sigma \rho(0)_{\beta\beta'}(j(\omega_\sigma - \omega_\beta) + ij'(\omega_\sigma - \omega_\beta))$$

$$\times \sum_{q=-j}^{j} A^{(q)}_{\sigma\beta} A^{(q)+}_{\alpha\sigma} \exp[i(\omega_\alpha - \omega_\beta + \omega_{\beta'} - \omega_{\alpha'})t]. \quad (3.55)$$

On substituting from (3.51), (3.52), (3.54) and (3.55) into (3.35) we deduce that

$$\frac{d\rho(t)^{\text{int}}_{\alpha\alpha'}}{dt} = \sum_{\beta\beta'} \left\{ (j(\omega_\alpha - \omega_\beta) + ij'(\omega_\alpha - \omega_\beta)) \sum_{q=-j}^{j} A^{(q)}_{\alpha\beta} A^{(q)+}_{\beta'\alpha'} \right.$$

$$+ (j(\omega_{\alpha'} - \omega_{\beta'}) + ij'(\omega_{\alpha'} - \omega_{\beta'})) \sum_{q=-j}^{j} A^{(q)}_{\alpha\beta} A^{(q)+}_{\beta'\alpha'}$$

$$- \delta_{\alpha\beta} \sum_\sigma (j(\omega_\sigma - \omega_{\beta'}) - ij'(\omega_\sigma - \omega_{\beta'})) \sum_{q=-j}^{j} A^{(q)}_{\sigma\alpha'} A^{(q)+}_{\beta'\sigma}$$

$$\left. - \delta_{\alpha'\beta'} \sum_\sigma (j(\omega_\sigma - \omega_\beta) + ij'(\omega_\sigma - \omega_\beta)) \sum_{q=-j}^{j} A^{(q)}_{\sigma\beta} A^{(q)+}_{\alpha\sigma} \right\}$$

$$\times \exp[i(\omega_\alpha - \omega_{\alpha'} - \omega_\beta + \omega_{\beta'})t]\rho(0)_{\beta\beta'}. \quad (3.56)$$

We may write this equation as

$$\frac{d\rho(t)^{\text{int}}_{\alpha\alpha'}}{dt} = \sum_{\beta\beta'} P_{\alpha\alpha'\beta\beta'} \exp[i(\omega_\alpha - \omega_{\alpha'} - \omega_\beta + \omega_{\beta'})t]\rho(0)_{\beta\beta'}, \quad (3.57)$$

and we see that $P_{\alpha\alpha'\beta\beta'}$ is time independent.

In deducing (3.29) from (3.27) we replaced $\rho(0)^{\text{int}}$ in the last term of (3.29) by $\rho(0)$ to which it is equal. We now write $\rho(0)_{\beta\beta'}$ in (3.57) as $\rho(0)^{\text{int}}_{\beta\beta'}$, and making the assumption that $\rho(0)^{\text{int}}_{\beta\beta'}$ may be replaced by $\rho(t)^{\text{int}}_{\beta\beta'}$ we obtain

$$\frac{d\rho(t)^{\text{int}}_{\alpha\alpha'}}{dt} = \sum_{\beta\beta'} P_{\alpha\alpha'\beta\beta'} \exp[i(\omega_\alpha - \omega_{\alpha'} - \omega_\beta + \omega_{\beta'})t]\rho(t)^{\text{int}}_{\beta\beta'}. \quad (3.58)$$

This agrees with the result of Redfield (1957) when we adopt his notation and neglect the imaginary $j'(\omega)$-terms in the definition of $P_{\alpha\alpha'\beta\beta'}$.

The structure of (3.58) resembles that of the equation

$$i\hbar \dot{b}_n(t) = \sum_m \exp[i(E_n - E_m)t/\hbar] b_m(t) \quad (3.59)$$

encountered in the quantum theory of perturbation (Heitler, 1954, section 14; McConnell, 1960, section 43). The equations differ in that, while (3.59) describes transitions between energy levels labelled m and n, (3.58) describes transitions between matrix elements labelled $\beta\beta'$ and $\alpha\alpha'$. Just as it is shown that an $m \to n$ transition occurs only if $E_n - E_m = 0$ in (3.59), it may also be shown (Redfield, 1957) that a condition for the relaxation process to occur is

$$\omega_\alpha - \omega_{\alpha'} - \omega_\beta + \omega_{\beta'} = 0. \tag{3.60}$$

It may nevertheless be mathematically convenient to retain the exponential term in (3.58).

Let us transform back from $\rho(t)^{\text{int}}$ to $\rho(t)$ in (3.58). From (3.33) we see that

$$\frac{d\rho(t)^{\text{int}}_{\alpha\alpha'}}{dt} = i(\omega_\alpha - \omega_{\alpha'})\rho(t)^{\text{int}}_{\alpha\alpha'} + \exp[i(\omega_\alpha - \omega_{\alpha'})t]\frac{d\rho(t)_{\alpha\alpha'}}{dt},$$

$$\rho(t)^{\text{int}}_{\beta\beta'} = \exp[i(\omega_\beta - \omega_{\beta'})t]\rho(t)_{\beta\beta'},$$

$$i[\rho'^+), \mathcal{H}_0]_{\alpha\alpha'} = i(\alpha|\rho(t)\mathcal{H}_0 - \mathcal{H}_0\rho(t)|\alpha') = i(\omega_{\alpha'} - \omega_\alpha)\rho(t)_{\alpha\alpha'}. \tag{3.61}$$

We then deduce from (3.58) that

$$\frac{d\rho(t)_{\alpha\alpha'}}{dt} = i[\rho(t), \mathcal{H}_0]_{\alpha\alpha'} + \sum_{\beta\beta'} P_{\alpha\alpha'\beta\beta'}\rho(t)_{\beta\beta'}. \tag{3.62}$$

The first term on the right hand side provides the change in $\rho(t)_{\alpha\alpha'}$ due to the constant field and the second term provides the change in $\rho(t)_{\alpha\alpha'}$ which comes from the perturbing field. $P_{\alpha\alpha'\beta\beta'}$ may be regarded as a transition probability from the density matrix element $\rho(t)_{\beta\beta'}$ to the element $\rho(t)_{\alpha\alpha'}$. The above assumption that we replace $\rho(0)^{\text{int}}_{\beta\beta'}$ by $\rho(t)^{\text{int}}_{\beta\beta'}$ supposes that we can find a value of t such that $tP_{\alpha\alpha'\beta\beta'} \ll 1$. Combining this with the earlier restriction on t we say that the conditions

$$\frac{1}{P_{\alpha\alpha'\beta\beta'}} \gg t \gg \tau_c \tag{3.63}$$

are necessary for the validity of the theory.

Equation (3.62) provides a convenient way of calculating the time rate of change of the ensemble average of a physical observable corresponding to the self-adjoint operator B. According to (3.6) and (3.7) the ensemble average

$$\langle B \rangle = \text{tr}(\rho B) = \sum_\alpha (\rho B)_{\alpha\alpha} = \sum_{\alpha\alpha'} \rho_{\alpha\alpha'} B_{\alpha'\alpha},$$

and so

$$\frac{d\langle B \rangle}{dt} = \sum_{\alpha\alpha'} \frac{d\rho(t)_{\alpha\alpha'}}{dt} B_{\alpha'\alpha}. \tag{3.64}$$

On substituting from (3.62) it is found that

$$\frac{d\langle B \rangle}{dt} = i\sum_{\alpha\alpha'}[\rho(t), \mathcal{H}_0]_{\alpha\alpha'}B_{\alpha'\alpha} + \sum_{\alpha\alpha'\beta\beta'} P_{\alpha\alpha'\beta\beta'}\rho(t)_{\beta\beta'}B_{\alpha'\alpha}. \tag{3.65}$$

3.4 Application to nuclear magnetic relaxation

In order to facilitate the application of results of the previous section to the various mechanisms of nuclear magnetic relaxation that will be studied in the following chapters we shall now discuss some general matters. It was explained in section 1.1 how relaxation occurs when a system of particles having spins interacts with both a strong constant field in the z-direction and a weak time dependent perturbing field. During the relaxation process the magnetization $\mathbf{M}$ per unit volume often obeys (1.20)–(1.22) for isotropic media.

Turning to nuclear magnetism we consider the case of identical nuclei. If the nuclear spin vector is $\mathbf{I}$, the gyromagnetic ratio is γ and the number of nuclei per unit volume is N, (1.6) gives

$$\mathbf{M} = N\gamma\hbar\langle \mathbf{I}\rangle. \tag{3.66}$$

We then see from (1.20)–(1.22) that the longitudinal and transverse relaxation times T_1, T_2 are obtained from

$$\frac{d\langle I_z\rangle}{dt} = -\frac{\langle I_z\rangle - \langle I_z\rangle_0}{T_1}, \tag{3.67}$$

$$\frac{d\langle I_x\rangle}{dt} = -\frac{\langle I_x\rangle}{T_2}, \quad \frac{d\langle I_y\rangle}{dt} = -\frac{\langle I_y\rangle}{T_2}, \tag{3.68}$$

where $\langle I_z\rangle_0$ is the value of $\langle I_z\rangle$ when equilibrium is reached. It will therefore suffice to consider $\langle I_r\rangle$, where $r = x$, y, z.

In the presence of the constant field H_0 in the z-direction the Hamiltonian for the nucleus $\hbar\mathcal{H}_0$ is given by

$$\hbar\mathcal{H}_0 = -\gamma\hbar I_z H_0 = -\hbar\omega_0 I_z, \tag{3.69}$$

where ω_0 is the Larmor angular frequency. When performing calculations for I_r, we shall see later that, when the value of $P_{\alpha\alpha'\beta\beta'}$ taken from (3.56) is substituted into (3.62), the $j(\omega_\alpha - \omega_\beta)$-, $j(\omega_{\alpha'} - \omega_{\beta'})$-, $j(\omega_\sigma - \omega_{\beta'})$-, $j(\omega_\sigma - \omega_\beta)$-terms will often produce an expression of the form

$$-[a_r j(0) + b_r j(\omega_0) + c_r j(2\omega_0)]\langle I_r\rangle,$$

where a_r, b_r, c_r are real non-negative constants. It follows that the j'-terms will give to the last term of (3.62) a contribution of the form

$$-i[a'_r j'(0) + b'_r j'(\omega_0) + c'_r j'(2\omega_0)] \sum_{\alpha\alpha'} \rho(t)_{\alpha\alpha'}(I_r)_{\alpha'\alpha}, \tag{3.70}$$

where a'_r, b'_r, c'_r are real constants. On the other hand we deduce from (3.61) that

$$i\sum_{\alpha\alpha'} [\rho(t), \mathcal{H}_0]_{\alpha\alpha'}(I_r)_{\alpha'\alpha} = i\sum_{\alpha\alpha'} (\omega_{\alpha'} - \omega_\alpha)\rho(t)_{\alpha\alpha'}(I_r)_{\alpha'\alpha},$$

which when added to (3.70) yields

$$i \sum_{\alpha\alpha'} \{ (\omega_{\alpha'} - \omega_\alpha) - [a'_r j'(0) + b'_r j'(\omega_0) + c'_r j'(2\omega_0)] \} \rho(t)_{\alpha\alpha'} (I_r)_{\alpha'\alpha}.$$

Hence the only effect of the j'-terms is to produce an additional small constant frequency shift.

We can take account of this shift by adding to the field H_0 a small constant term in addition to the small term already added in the course of the derivation of (3.30). Denoting the new value of the constant field by H'_0 we shall put $\mathcal{H}'_0 = -\gamma H'_0 I_z$ and in future we shall interpret ω_0 as $\gamma H'_0$. We then replace (3.65) by

$$\frac{d\langle B \rangle}{dt} = i \sum_{\alpha\alpha'} [\rho(t), \mathcal{H}'_0]_{\alpha\alpha'} B_{\alpha'\alpha} + \sum_{\alpha'\beta\beta'} P'_{\alpha\alpha'\beta\beta'} \rho(t)_{\beta\beta'} B_{\alpha'\alpha}, \qquad (3.71)$$

where $P'_{\alpha\alpha'\beta\beta'}$ is obtained from (3.56) and (3.57) by omitting the j'-terms. The eigenfunctions of $h\mathcal{H}'_0$ are now used for the basis of the representation. This will produce a small change in the values of the ω_α's but we may neglect the influence of this change in the last term of (3.71) which is already small in comparison with the previous term. The relaxation process arises from $hG(t)$ and hence for relaxation (3.71) reduces, for $B \equiv I_r$, to

$$\frac{d\langle I_r \rangle}{dt} = \sum_{\alpha\alpha'\beta\beta'} \left\{ j(\omega_\alpha - \omega_\beta) \sum_{q=-j}^{j} A_{\alpha\beta}^{(q)} A_{\beta'\alpha'}^{(q)+} + j(\omega_{\alpha'} - \omega_{\beta'}) \sum_{q=-j}^{j} A_{\alpha\beta}^{(q)} A_{\beta'\alpha'}^{(q)+} \right.$$

$$\left. - \delta_{\alpha\beta} \sum_\sigma j(\omega_\sigma - \omega_{\beta'}) \sum_{q=-j}^{j} A_{\sigma\alpha'}^{(q)} A_{\beta'\sigma}^{(q)+} - \delta_{\alpha'\beta'} \sum_\sigma j(\omega_\sigma - \omega_\beta) \sum_{q=-j}^{j} A_{\sigma\beta}^{(q)} A_{\alpha\sigma}^{(q)+} \right\}$$

$$\times \rho(t)_{\beta\beta'} (I_r)_{\alpha'\alpha}, \qquad (3.72)$$

where $j(\omega)$, $A^{(q)}$, $A^{(q)+}$ are to be found from (3.38) and (3.48) and we have employed (3.60).

Equation (3.72) is the most convenient one for computational purposes. An alternative form obtained from (3.49) is

$$\frac{d\langle I_r \rangle}{dt} = \sum_{\alpha\alpha'\beta\beta'} R_{\alpha\alpha'\beta\beta'} \rho(t)_{\beta\beta'} (I_r)_{\alpha'\alpha} \qquad (3.73)$$

with

$$R_{\alpha\alpha'\beta\beta'} = J_{\alpha\beta\alpha'\beta'}(\omega_\alpha - \omega_\beta) + J_{\alpha\beta\alpha'\beta'}(\omega_{\alpha'} - \omega_{\beta'})$$

$$- \delta_{\alpha\beta} \sum_\sigma J_{\sigma\alpha'\sigma\beta'}(\omega_\sigma - \omega_{\beta'}) - \delta_{\alpha'\beta'} \sum_\sigma J_{\sigma\beta\sigma\alpha}(\omega_\sigma - \omega_\beta).$$

Equation (3.73) could be deduced from (3.64) by employing the density matrix satisfying

$$\frac{d\rho(t)_{\alpha\alpha'}}{dt} = \sum_{\beta\beta'} R_{\alpha\alpha'\beta\beta'} \rho(t)_{\beta\beta'}. \qquad (3.74)$$

For relaxation processes this equation is equivalent to the Redfield result quoted after (3.58).

When (3.72) leads to (3.67) and (3.68), the relaxation times exist. Moreover it is not difficult to verify that

$$i \sum_{\alpha\alpha'} [\rho, \mathcal{H}_0]_{\alpha\alpha'} \mathbf{M}_{\alpha'\alpha} = \gamma(\mathbf{M} \times \mathbf{H}_0)$$

for a constant external field H_0 in a fixed direction which produces an interaction Hamiltonian $\hbar\mathcal{H}_0$. On employing (3.71) with $B \equiv \mathbf{M}$ we may then obtain the Bloch equations (1.24).

Let us consider the conditions for the applicability of the foregoing theory, apart from those stated in (3.63). Of the various nuclear magnetic relaxation processes mentioned in section 1.4, all except scalar relaxation of the second kind, to be described in section 7.3, have a rotationally invariant interaction Hamiltonian $\hbar G(t)$, where $G(t)$ is given by (3.38) and $H_q(t)$ is a spin independent commuting function. With this single exception the theory is applicable to all the relaxation processes, be they intermolecular or intramolecular.

In order to employ (3.72) we need to know the spectral density $j(\omega)$ defined in (3.48). Of the various relaxation processes to which the theory is applicable, all except intermolecular dipole–dipole interactions and spin-rotational interactions have the property that $H_q(t)$ is a function of the angular variables only. Assuming that $H_q(t)$ is such a function we deduce from (3.48) and (C.33) that

$$j(\omega) = \frac{1}{2(2j+1)} \sum_{m,m'=-j}^{j} H_m'^* H_m' \int_{-\infty}^{\infty} \langle R^+(t)\rangle_{mm'} \, e^{-i\omega t} \, dt. \qquad (3.75)$$

In this equation H_p' is the constant pth spherical component in a body frame of coordinates. $R(t)$ is the rotation operator that corresponds to the rotation of the molecule in question from its orientation at time zero to its orientation at time t. The subscripts mm' denote the mm'-matrix element in the representation with basis $Y_{j,-j}, Y_{j,-j+1}, \ldots, Y_{j,j+1}, Y_{jj}$, the spherical harmonics being associated with the body frame.

The integral in (3.75) depends only on the random rotation of the molecule and is therefore independent of the particular relaxation process that is being investigated. For different processes the value of H_p' will be different. Then when we have calculated the integral for a certain model of the molecule that contains the nucleus in which we are interested, e.g. for a spherical, linear, asymmetric molecular model, we can employ the value of the integral for all relaxation processes whose $H_q(t)$ is a function of the angular variables only.

4

Dipolar interactions

4.1 The dipole–dipole interaction

A mechanism that one would expect to cause nuclear magnetic relaxation is the interaction of the spin of the nucleus in which we are interested with spins in the same molecule or in neighbouring molecules (McConnell, 1986a). To study such an interaction we begin with the standard result in magnetostatics that the mutual potential energy Φ of two magnets with moments μ, μ' is given by (Jeans, 1933, p. 378)

$$\Phi = \frac{(\mu \cdot \mu')}{r^3} - \frac{3(\mu \cdot \mathbf{r})(\mu' \cdot \mathbf{r})}{r^5}, \tag{4.1}$$

where $\mathbf{r}$ is the position vector of the magnet with moment μ' relative to the other. If we take a set of cartesian axes with origin at the latter magnet and in directions fixed with respect to the laboratory, and if we denote by (x, y, z) the components of $\mathbf{r}$, we find from (4.1) that

$$\Phi = \frac{1}{r^3}\left[\left(1 - \frac{3x^2}{r^2}\right)\mu_x\mu'_x + \left(1 - \frac{3y^2}{r^2}\right)\mu_y\mu'_y + \left(1 - \frac{3z^2}{r^2}\right)\mu_z\mu'_z\right.$$

$$-\frac{3yz}{r^2}(\mu_y\mu'_z + \mu_z\mu'_y) - \frac{3zx}{r^2}(\mu_z\mu'_x + \mu_x\mu'_z)$$

$$\left. -\frac{3xy}{r^2}(\mu_x\mu'_y + \mu_y\mu'_x)\right]. \tag{4.2}$$

On introducing the spherical polar angles θ, ϕ by

$$x = r \sin\theta\cos\phi, \qquad y = r\sin\theta\sin\phi, \qquad z = r\cos\theta \tag{4.3}$$

we deduce that

$$r^3\Phi = (1 - 3\sin^2\theta\cos^2\phi)\mu_x\mu'_x + (1 - 3\sin^2\theta\sin^2\phi)\mu_y\mu'_y$$

$$+ (1 - 3\cos^2\theta)\mu_z\mu'_z - 3\sin\theta\cos\theta\sin\phi(\mu_y\mu'_z + \mu_z\mu'_y)$$

$$- 3\sin\theta\cos\theta\cos\phi(\mu_z\mu'_x + \mu_x\mu'_z)$$

$$- 3\sin^2\theta\sin\phi\cos\phi(\mu_x\mu'_y + \mu_y\mu'_x). \tag{4.4}$$

From the definition of spherical harmonics given in (B.1) of Appendix B we obtain as in (B.28)

$$Y_{1,-1}(\theta, \phi) = \left(\frac{3}{8\pi}\right)^{1/2} (\cos\phi - i\sin\phi)\sin\theta,$$

$$Y_{10}(\theta, \phi) = \left(\frac{3}{4\pi}\right)^{1/2} \cos\theta, \tag{4.5}$$

$$Y_{11}(\theta, \phi) = -\left(\frac{3}{8\pi}\right)^{1/2} (\cos\phi + i\sin\phi)\sin\theta.$$

If θ_1, ϕ_1 are the polar angles of the axis of the magnet with moment μ, so that

$$\mu_x = \mu\sin\theta_1\cos\phi_1, \qquad \mu_y = \mu\sin\theta_1\sin\phi_1, \qquad \mu_z = \mu\cos\theta_1,$$

we see from (4.5) that μ_{-1}, μ_0, μ_1 defined by

$$\mu_{-1} = \frac{\mu_x - i\mu_y}{2^{1/2}}, \qquad \mu_0 = \mu_z, \qquad \mu_1 = -\frac{\mu_x + i\mu_y}{2^{1/2}} \tag{4.6}$$

transform like the spherical harmonics. Thus μ_{-1}, μ_0, μ_1 are the components of a spherical tensor of rank 1 as defined in Appendix B. The same is true for $\mu'_{-1}, \mu'_0, \mu'_1$ given by equations analogous to (4.6). We may express the cartesian components of μ and μ' in terms of their spherical components by

$$\mu_x = \frac{\mu_{-1} - \mu_1}{2^{1/2}}, \qquad \mu_y = \frac{i(\mu_{-1} + \mu_1)}{2^{1/2}}, \qquad \mu_z = \mu_0$$

$$\mu'_x = \frac{\mu'_{-1} - \mu'_1}{2^{1/2}}, \qquad \mu'_y = \frac{i(\mu'_{-1} + \mu'_1)}{2^{1/2}}, \qquad \mu'_z = \mu'_0.$$

On substituting from these equations into (4.4) we find that

$$r^3\Phi = \tfrac{1}{2}[3\mu_0\mu'_0 - (\boldsymbol{\mu}\cdot\boldsymbol{\mu}')](1 - 3\cos^2\theta) - \frac{3}{2^{1/2}}(\mu_{-1}\mu'_0 + \mu_0\mu'_{-1})\sin\theta\cos\theta\, e^{i\phi}$$

$$+ \frac{3}{2^{1/2}}(\mu_1\mu'_0 + \mu_0\mu'_1)\sin\theta\cos\theta\, e^{-i\phi} - \tfrac{3}{2}\mu_1\mu'_1\sin^2\theta\, e^{-2i\phi}$$

$$- \tfrac{3}{2}\mu_{-1}\mu'_{-1}\sin^2\theta\, e^{2i\phi}. \tag{4.7}$$

From (B.30) we also have

$$Y_{20}(\theta, \phi) = \left(\frac{5}{16\pi}\right)^{1/2} (3\cos^2\theta - 1),$$

$$Y_{2,\pm1}(\theta, \phi) = \mp\left(\frac{15}{8\pi}\right)^{1/2} \cos\theta\sin\theta\, e^{\pm i\phi}, \tag{4.8}$$

$$Y_{2,\pm2}(\theta, \phi) = \left(\frac{15}{32\pi}\right)^{1/2} \sin^2\theta\, e^{\pm 2i\phi},$$

and these equations allow us to express (4.7) as

$$\Phi = \left(\frac{4\pi}{5}\right)^{1/2} \frac{1}{r^3} \{ -Y_{20}(\theta,\phi)[3\mu_0\mu_0' - (\boldsymbol{\mu}\cdot\boldsymbol{\mu}')]$$
$$+ 3^{1/2}Y_{21}(\theta,\phi)(\mu_{-1}\mu_0' + \mu_0\mu_{-1}')$$
$$+ 3^{1/2}Y_{2,-1}(\theta,\phi)(\mu_1\mu_0' + \mu_0\mu_1') - 6^{1/2}Y_{22}(\theta,\phi)\mu_{-1}\mu_{-1}'$$
$$- 6^{1/2}Y_{2,-2}(\theta,\phi)\mu_1\mu_1'\}. \tag{4.9}$$

It is shown in Appendix C that we may construct from $\boldsymbol{\mu}$ and $\boldsymbol{\mu}'$ a spherical tensor V of rank 2 whose five components are given by

$$V_0 = \frac{1}{6^{1/2}} [3\mu_0\mu_0' - (\boldsymbol{\mu}\cdot\boldsymbol{\mu}')]$$

$$V_{\pm 1} = \frac{1}{2^{1/2}} (\mu_0\mu_{\pm 1}' + \mu_{\pm 1}\mu_0') \tag{4.10}$$

$$V_{\pm 2} = \mu_{\pm 1}\mu_{\pm 1}'.$$

On substituting (4.10) into (4.9) we have

$$\Phi = -\left(\frac{24\pi}{5}\right)^{1/2} \frac{1}{r^3} [Y_{20}(\theta,\phi)V_0 - Y_{21}(\theta,\phi)V_{-1} - Y_{2,-1}(\theta,\phi)V_1$$
$$+ Y_{22}(\theta,\phi)V_{-2} + Y_{2,-2}(\theta,\phi)V_2],$$

so that

$$\Phi = -\left(\frac{24\pi}{5}\right)^{1/2} \sum_{m=-2}^{2} (-)^m r^{-3} Y_{2m}(\theta,\phi)V_{-m}. \tag{4.11}$$

The calculations have so far been based entirely on classical electromagnetic theory. We now relate the magnetic moments $\boldsymbol{\mu}$, $\boldsymbol{\mu}'$ to two independent particles which are usually two atomic nuclei but may also be a nucleus and an electron. We denote the quantum mechanical spins of the particles by the Hermitian operators $\mathbf{I}$ and $\mathbf{S}$, and the respective gyromagnetic ratios by γ_I and γ_S. In accordance with (1.6) we put

$$\boldsymbol{\mu} = \gamma_I \hbar \mathbf{I}, \qquad \boldsymbol{\mu}' = \gamma_S \hbar \mathbf{S}. \tag{4.12}$$

To conform with the customary notation for raising and lowering operators we write

$$\begin{aligned} I_+ &= I_x + iI_y, & I_- &= I_x - iI_y, & I_0 &= I_z \\ S_+ &= S_x + iS_y, & S_- &= S_x - iS_y, & S_0 &= S_z, \end{aligned} \tag{4.13}$$

from which it is immediately seen that

$$I_+^+ = I_-, \qquad S_+^+ = S_-. \tag{4.14}$$

From (4.6) and (4.13)

$$\mu_{-1}=\frac{\gamma_I\hbar I_-}{2^{1/2}}, \qquad \mu_0=\gamma_I\hbar I_z, \qquad \mu_1=-\frac{\gamma_I\hbar I_+}{2^{1/2}}$$

$$\mu'_{-1}=\frac{\gamma_S\hbar S_-}{2^{1/2}}, \qquad \mu'_0=\gamma_S\hbar S_z, \qquad \mu'_1=-\frac{\gamma_S\hbar S_+}{2^{1/2}}. \qquad (4.15)$$

When (4.15) are substituted into (4.10), we obtain

$$V_0=\frac{\gamma_I\gamma_S\hbar^2}{6^{1/2}}[3I_0S_0-(\mathbf{I}\cdot\mathbf{S})]$$

$$V_{\pm1}=\mp\frac{\gamma_I\gamma_S\hbar^2}{2}[I_0S_\pm+I_\pm S_0] \qquad (4.16)$$

$$V_{\pm2}=\frac{\gamma_I\gamma_S\hbar^2}{2}I_\pm S_\pm.$$

Equation (4.11) still holds with V_{-m} defined by (4.16). The V's are now quantum mechanical operators and are therefore in general non-commuting quantities. However, since $\mathbf{I}$ and $\mathbf{S}$ refer to two independent particles, a component of $\mathbf{I}$ will commute with a component of $\mathbf{S}$. Constructing a classical interaction potential and then interpreting variables that it contains as operators is a procedure frequently adopted in quantum field theory calculations. We see from (4.2) that for random motion the average value of Φ is zero and therefore that the interaction does not produce a shift of energy levels.

Let us express the above results in a form suitable for the study of nuclear magnetic relaxation by the interaction of a pair of dipoles at a distance r apart, when the two-dipole system is subject to random thermal motion. We assume that the static Φ remains a good approximation for the interaction potential, and to conform with the notation of (3.16) we replace Φ by $\hbar G(t)$, the time dependence resulting from the random motion which makes r and the angles θ, ϕ functions of t. We then express (4.11) as

$$\hbar G(t)=-\left(\frac{6\pi}{5}\right)^{1/2}\gamma_I\gamma_S\hbar^2\sum_{q=-2}^{2}(-)^q r(t)^{-3}Y_{2,-q}(\theta(t),\phi(t))A^{(q)}, \quad (4.17)$$

where, from (4.16),

$$A^{(0)}=\frac{2}{6^{1/2}}[3I_0S_0-(\mathbf{I}\cdot\mathbf{S})]$$

$$A^{(-1)}=I_0S_-+I_-S_0, \qquad A^{(1)}=-(I_0S_++I_+S_0) \qquad (4.18)$$

$$A^{(-2)}=I_-S_-, \qquad A^{(2)}=I_+S_+.$$

It is immediately verified from (4.8), (4.17) and (4.18) that $\hbar G(t)$ is a self-adjoint operator. The spin dependence of the interaction Hamiltonian resides entirely in $A^{(q)}$. Equation (4.17) is expressible in the form (3.38) with $j = 2$ and

$$H_q(t) = -\left(\frac{6\pi}{5}\right)^{1/2} \gamma_I \gamma_S \hbar r(t)^{-3} Y_{2q}(\theta(t), \phi(t)), \qquad (4.19)$$

which shows that $H_q(t)$ is a component of a spherical tensor of rank 2. Then, from (3.48), we have the spectral density

$$j(\omega) = \frac{3\pi}{5} \gamma_I^2 \gamma_S^2 \hbar^2 \int_{-\infty}^{\infty} \langle r(0)^{-3} Y_{20}(\theta(0), \phi(0)) r(t)^{-3} Y_{20}(\theta(t), \phi(t)) \rangle \, e^{-i\omega t} \, dt. \qquad (4.20)$$

We may use (2.16) to express this as

$$j(\omega) = \frac{3\pi \gamma_I^2 \gamma_S^2 \hbar^2}{5} \int_{-\infty}^{\infty} dt \, e^{-i\omega t} \iint p(\mathbf{r}(0), 0) w(\mathbf{r}(0), 0; \mathbf{r}(t), t)$$
$$\times r(0)^{-3} Y_{20}(\theta(0), \phi(0)) r(t)^{-3} Y_{20}(\theta(t), \phi(t)) \, d^3\mathbf{r}(0) \, d^3\mathbf{r}(t). \qquad (4.21)$$

Equations (4.20) and (4.21) are true for both translational and rotational thermal motion, and for both intermolecular and intramolecular dipolar interactions. If we are concerned only with intramolecular interactions and if we regard the molecules as rigid, $r(0)$ and $r(t)$ have a common value r, say. Equation (4.20) then simplifies to

$$j(\omega) = \frac{3\pi \gamma_I^2 \gamma_S^2 \hbar^2}{5r^6} \int_{-\infty}^{\infty} \langle Y_{20}(\theta(0), \phi(0)) Y_{20}(\theta(t), \phi(t)) \rangle \, e^{-i\omega t} \, dt. \qquad (4.22)$$

Referring to Appendix C we see from (C.31) and (4.22) that

$$j(\omega) = \frac{3\pi \gamma_I^2 \gamma_S^2 \hbar^2}{25r^6} \sum_{m,m'=-2}^{2} Y_{2m'}^*(\theta', \phi') Y_{2m}(\theta', \phi') \int_{-\infty}^{\infty} \langle R^+(t) \rangle_{mm'} \, e^{-i\omega t} \, dt. \qquad (4.23)$$

In (4.23) the angles θ', ϕ' determine the orientation of the dipole–dipole axis with respect to the body coordinate frame introduced in Appendix C.

We shall in the following sections present a relaxation theory for dipolar interactions at some length, so that it may serve as a model for later calculations arising from other interactions. In this chapter we concentrate on deriving relations that express the relaxation times T_1, T_2 in terms of $j(\omega)$. In Chapters 5 and 6 we shall again in some detail work out the consequences of these relations for inter- and intramolecular interactions.

4.2 Systems of like spins

We apply results of the previous section to magnetic material in the liquid

state which is undergoing random motion. Let us take the case of a
molecule which contains two nuclei of like spin, for example, the proton
nuclei in the water molecule H_2O. We shall now make a rough estimate of
the flux density B caused by the field that one proton exerts on the other.
According to the previous section this would be of order μr^{-3}, where r is
the distance of one proton from the other. To allow the use of SI units we
convert μr^{-3} to $\mu_0 \mu_p / 4\pi r^3$, where the permeability of free space $\mu_0 = 4\pi \times 10^{-7}$ H m^{-1}, the magnetic moment of the proton $\mu_p = 1.521 \times 10^{-26}$ J T^{-1}
and $r = 1.5 \times 10^{-10}$ m (Herzberg, 1945, p. 489). Thus we find

$$B = \frac{4\pi \times 1.521 \times 10^{30}}{4\pi \times 1.5^3 \times 10^{33}} \text{ H m}^{-4} \text{ J T}^{-1} = 0.451 \times 10^{-3} \text{ T}.$$

On the other hand the value of B for the constant laboratory field is of order
0.1–1 T. We are therefore justified in regarding the dipole–dipole
interaction as a small perturbation of the laboratory field. A similar
argument will clearly be valid for two interacting nuclei in any molecule
whether their spins are like or unlike, and *a fortiori* it will be valid for two
interacting nuclei in different molecules, since B decreases with r^{-3}.

We proceed to examine the general case of two like spins putting
$\gamma_I = \gamma_S = \gamma$, so that (4.19) becomes

$$H_q(t) = -\left(\frac{6\pi}{5}\right)^{1/2} \frac{\gamma^2 \hbar}{r(t)^3} Y_{2q}(\theta(t), \phi(t)). \tag{4.24}$$

Though the spins I and S are equal, the operators $\mathbf{I}$ and $\mathbf{S}$ refer to different
particles and so each component of $\mathbf{I}$ commutes with every component of $\mathbf{S}$.
Employing the expression $-(\boldsymbol{\mu} \cdot \mathbf{H})$ for the energy of a magnet with moment
$\boldsymbol{\mu}$ in the presence of a field $\mathbf{H}$ we see that the Hamiltonian for the magnets of
moments $\gamma\hbar\mathbf{I}$ and $\gamma\hbar\mathbf{S}$ in the presence of a d.c. magnetic field H_0 in the
positive z-direction is $(-\gamma\hbar I_z - \gamma\hbar S_z)H_0$. Replacing γH_0 by ω_0, as in (1.10),
we express the unperturbed Hamiltonian $\hbar\mathcal{H}_0$ of (3.16) by

$$\hbar\mathcal{H}_0 = -\hbar\omega_0(I_z + S_z). \tag{4.25}$$

Let us apply Redfield's equations to the two-spin system. We first
examine the arguments of the j-functions. When $q = 0$ in the summations,
$A^{(q)}$ is given from (4.18) by

$$A^{(0)} = \frac{2}{6^{1/2}} [2I_z S_z - I_x S_x - I_y S_y], \tag{4.26}$$

which shows that $A^{(0)}$ is self-adjoint. We then see from (4.13) that

$$A^{(0)} = \frac{2}{6^{1/2}} [2I_0 S_0 - \tfrac{1}{2}(I_+ S_- + I_- S_+)]. \tag{4.27}$$

Consider now the term

$$j(\omega_\alpha - \omega_\beta)(\alpha|A^{(0)}|\beta)(\beta'|A^{(0)}|\alpha') \tag{4.28}$$

in (3.72). The $|\alpha)$, $|\beta)$ etc. are eigenfunctions of the unperturbed Hamiltonian $\hbar\mathcal{H}_0$ of (4.25), and these are sums of products of the eigenfunctions of I_z and S_z. If $|m_I)$ is a normalized eigenfunction of I_z with eigenvalue m_I and $|m_S)$ is a normalized eigenfunction of S_z with eigenvalue m_S, then $|m_I)|m_S)$ is an eigenfunction of $I_z + S_z$ with eigenvalue $m_I + m_S$. The corresponding value of the unperturbed energy is $-(m_I + m_S)\hbar\omega_0$. Thus, if $|\beta)$ denotes $|m_I)|m_S)$, it follows that $(\alpha|I_0 S_0|\beta)$ will vanish unless $|\alpha)$ also denotes $|m_I)|m_S)$. The energies of the states $|\alpha)$ and $|\beta)$ will therefore be equal, and denoting the energies by $\hbar\omega_\alpha$ and $\hbar\omega_\beta$, as in section 3.3, $\omega_\alpha - \omega_\beta$ will be equal to zero.

Turning to the other operators in (4.27) we know that, since I_+ and I_- are, respectively, raising and lowering operators,

$$I_+|m_I) = a_I|m_I + 1), \qquad I_-|m_I) = b_I|m_I - 1),$$

where a_I, b_I are constants that may be zero. Similarly we have

$$S_+|m_S) = a_S|m_S + 1), \qquad S_-|m_S) = b_S|m_S - 1),$$

and thus

$$I_+ S_-|m_I)|m_S) = a_I b_S|m_I + 1)|m_S - 1). \tag{4.29}$$

Hence $(\alpha|I_+ S_-|\beta)$ will vanish except when $|\alpha)$ denotes $|m_I + 1)|m_S - 1)$. Then

$$\hbar\omega_\alpha = -\hbar\omega_0(m_I + 1 + m_S - 1) = \hbar\omega_\beta,$$

so that $\omega_\alpha - \omega_\beta$ vanishes. The same will be true for $(\alpha|I_- S_+|\beta)$. We therefore see that, when $q = 0$, the first term on the right hand side of (3.72) is proportional to $j(0)$. The same result will clearly follow for the second term.

The above reasoning when applied to $A^{(-1)}$ and $A^{(1)}$ as defined in (4.18) will immediately show that $(\alpha|A^{(\pm 1)}|\beta)$ vanishes unless

$$\hbar\omega_\alpha - \hbar\omega_\beta = \pm\hbar\omega_0.$$

Since $j(\omega)$ is an even function, it follows that the first two terms on the right hand side of (3.72) with $q = 1$ will be multiplied by $j(\omega_0)$. Similarly the terms with $q = 2$ will be multiplied by $j(2\omega_0)$. The same line of argument is applicable to the double summation terms on the right hand side of (3.72).

We now put

$$B = I_r + S_r \tag{4.30}$$

in (3.71), so that I_r is replaced by $I_r + S_r$ in (3.72). Since $A^{(0)+} = A^{(0)}$, the coefficient of $j(0)$ is

$$\sum_{\alpha\alpha'\beta\beta'} \rho(t)_{\beta\beta'} [2A^{(0)}_{\beta'\alpha'}(I_r+S_r)_{\alpha'\alpha}A^{(0)}_{\alpha\beta} - \sum_{\sigma=-2}^{2} A^{(0)}_{\beta'\sigma}A^{(0)}_{\sigma\alpha'}(I_r+S_r)_{\alpha'\alpha}\delta_{\alpha\beta}$$

$$- \sum_{\sigma=-2}^{2} \delta_{\beta'\alpha'}(I_r+S_r)_{\alpha'\alpha}A^{(0)}_{\alpha\sigma}A^{(0)}_{\sigma\beta}].$$

On summing over $\alpha\alpha'\beta\beta'$ we obtain

$$\mathrm{tr}[\rho(t)\{2A^{(0)}(I_r+S_r)A^{(0)} - (A^{(0)})^2(I_r+S_r) - (I_r+S_r)(A^{(0)})^2\}],$$

which is equal to the ensemble average

$$\langle 2A^{(0)}(I_r+S_r)A^{(0)} - (A^{(0)})^2(I_r+S_r) - (I_r+S_r)(A^{(0)})^2 \rangle.$$

Expressing the quantity inside the angular brackets as a double commutator we find that the coefficient of $j(0)$ on the right hand side of (3.72) is

$$\langle [[A^{(0)}, I_r+S_r], A^{(0)}]\rangle. \tag{4.31}$$

The coefficient of $j(\omega_0)$ in (3.72) for B given by (4.30) is obtained from $q=\pm 1$. Proceeding as for $q=0$ we obtain

$$\langle [[A^{(1)}, [A^{(-1)}, I_r+S_r]] + [A^{(-1)}, [A^{(1)}, I_r+S_r]]\rangle. \tag{4.32}$$

Since in general

$$[A, [B, C]]^+ = (ABC - ACB - BCA + CBA)^+$$

$$= C^+B^+A^+ - B^+C^+A^+ - A^+C^+B^+ + A^+B^+C^+$$

$$= [A^+, [B^+, C^+]], \tag{4.33}$$

we may express (4.32) as

$$\langle [A^{(1)}, [A^{(-1)}, I_r+S_r]]\rangle + \mathrm{adj}, \tag{4.34}$$

where adj denotes adjoint operator. It is deduced in like manner that the coefficient of $j(2\omega_0)$ is

$$\langle [[A^{(-2)}, I_r+S_r], A^{(2)}]\rangle + \mathrm{adj}. \tag{4.35}$$

In performing detailed calculations the following relations are useful:

$$[I_y, I_z] = iI_x, \qquad [I_z, I_x] = iI_y, \qquad [I_x, I_y] = iI_z$$
$$[I_+, I_x] = I_z, \qquad [I_+, I_y] = iI_z, \qquad [I_+, I_z] = -I_+$$
$$[I_-, I_x] = -I_z, \qquad [I_-, I_y] = iI_z, \qquad [I_-, I_z] = I_-$$
$$[I_+, I_-] = 2I_z. \tag{4.36}$$

Similar equations hold for S_x, S_y, S_z, S_+, S_-. When B and D are independent of A and C, we also have

$$[AB, CD] = AC[B, D] + [A, C]DB. \tag{4.37}$$

Let us calculate (4.31). Substituting for $A^{(0)}$ from (4.26) we have

$$\tfrac{2}{3}\langle [[2I_zS_z - I_xS_x - I_yS_y, I_r+S_r], 2I_zS_z - I_xS_x - I_yS_y]\rangle. \tag{4.38}$$

On putting $r = z$ and employing (4.36) it is easily deduced that (4.38) and therefore (4.31) vanishes, and so the coefficient of $j(0)$ is zero. When $r = x$, we have

$$[2I_zS_z - I_xS_x - I_yS_y, I_x + S_x] = 3i(I_yS_z + I_zS_y),$$

so that

$$(4.31) = 2i\langle[I_yS_z + I_zS_y, 2I_zS_z - I_xS_x - I_yS_y]\rangle.$$

By using (4.37) it is found after an elementary calculation that

$$(4.31) = \langle(-4I_xS_z^2 - 4I_z^2S_x + 2I_yI_xS_y - 2I_zS_xS_z$$
$$-2I_zI_xS_z + 2I_yS_xS_y - 2I_y^2S_x - 2I_xS_y^2)\rangle.$$

On account of the mutual independence of I and S we may put

$$\langle[[A^{(0)}, I_x + S_x], A^{(0)}]\rangle$$
$$= -\langle I_x\rangle\langle2S_y^2 + 4S_z^2\rangle - \langle2I_y^2 + 4I_z^2\rangle\langle S_x\rangle \qquad (4.39)$$
$$+2\langle I_yI_x\rangle\langle S_y\rangle + 2\langle I_y\rangle\langle S_xS_y\rangle - 2\langle I_z\rangle\langle S_xS_z\rangle$$
$$-2\langle I_zI_x\rangle\langle S_z\rangle.$$

For $r = z$ in (4.34) we have from (4.18) and (4.36)

$$[A^{(-1)}, I_z + S_z] = [I_zS_- + I_-S_z, I_z + S_z]$$
$$= I_zS_- + I_-S_z = A^{(-1)},$$

so that

$$[A^{(1)}, [A^{(-1)}, I_z + S_z]] = [A^{(1)}, A^{(-1)}],$$

which is self-adjoint by (4.14) and (4.18). Hence the value of (4.34) with $r = z$ is $2[A^{(1)}, A^{(-1)}]$. Employing (4.36) and (4.37) we find from (4.34) that the coefficient of $j(\omega_0)$ is

$$-4\langle I_z^2\rangle\langle S_z\rangle - 4\langle I_z\rangle\langle S_z^2\rangle + 2\langle I_zI_+\rangle\langle S_-\rangle$$
$$+2\langle I_+\rangle\langle S_zS_-\rangle + 2\langle I_-I_z\rangle\langle S_+\rangle + 2\langle I_-\rangle\langle S_+S_z\rangle.$$

This is expressible as

$$-4\langle I_z^2\rangle\langle S_z\rangle - 4\langle I_z\rangle\langle S_z^2\rangle + 2\langle I_x\rangle\langle S_zS_x + S_xS_z\rangle$$
$$+2\langle I_y\rangle\langle S_yS_z + S_zS_y\rangle + 2\langle I_zI_x + I_xI_z\rangle\langle S_x\rangle + 2\langle I_yI_z + I_zI_y\rangle\langle S_y\rangle. \qquad (4.40)$$

When $r = x$ in (4.30) it is found in a similar manner that the coefficient of $j(\omega_0)$ in (3.72) is

$$\langle I_x\rangle\langle -2S_x^2 - 4S_y^2 - 4S_z^2 - S_z\rangle + \langle -2I_z^2 - 4I_x^2 - 4I_z^2 + I_z\rangle\langle S_x\rangle$$
$$+ \langle I_z\rangle\langle S_zS_x + S_xS_z\rangle + \langle I_zI_x + I_xI_z\rangle\langle S_z\rangle$$
$$-i\langle I_+\rangle\langle S_xS_y\rangle - i\langle I_yI_x\rangle\langle S_+\rangle + i\langle I_-\rangle\langle S_yS_x\rangle + i\langle I_xI_y\rangle\langle S_-\rangle.$$

This simplifies to

$$-\langle I_x\rangle\langle 2S_x^2+4S_y^2+4S_z^2\rangle-\langle 2I_x^2+4I_y^2+4I_z^2\rangle\langle S_x\rangle$$
$$+\langle I_y\rangle\langle S_xS_y+S_yS_x\rangle+\langle I_xI_y+I_yI_x\rangle\langle S_y\rangle \qquad (4.41)$$
$$+\langle I_z\rangle\langle S_zS_x+S_xS_z\rangle+\langle I_zI_x+I_xI_z\rangle\langle S_z\rangle.$$

The calculations arising from (4.35) are less complicated. It is found for r equal to z that the coefficient of $j(2\omega_0)$ in (3.72) is

$$-8\langle I_z\rangle\langle S_x^2+S_y^2\rangle-8\langle I_x^2+I_y^2\rangle\langle S_z\rangle, \qquad (4.42)$$

and that for r equal to x the coefficient of $j(2\omega_0)$ is

$$-2\langle I_x\rangle\langle S_x^2+S_y^2\rangle-2\langle I_x^2+I_y^2\rangle\langle S_x\rangle$$
$$+2\langle I_zI_x+I_xI_z\rangle\langle S_z\rangle+2\langle I_z\rangle\langle S_zS_x+S_xS_z\rangle. \qquad (4.43)$$

On collecting the results from (4.31), (4.34), (4.35) and (4.39)–(4.43) we deduce that

$$\frac{d\langle I_z+S_z\rangle}{dt}=-[4\langle I_z\rangle\langle S_z^2\rangle+4\langle I_z^2\rangle\langle S_z\rangle-2\langle I_x\rangle\langle S_zS_x+S_xS_z\rangle$$

$$-2\langle I_y\rangle\langle S_yS_z+S_zS_y\rangle-2\langle I_zI_x+I_xI_z\rangle\langle S_x\rangle$$
$$-2\langle I_yI_z+I_zI_y\rangle\langle S_y\rangle]j(\omega_0)$$
$$-[8\langle I_z\rangle\langle S_x^2+S_y^2\rangle+8\langle I_x^2+I_y^2\rangle\langle S_z\rangle]j(2\omega_0), \qquad (4.44)$$

$$\frac{d\langle I_x+S_x\rangle}{dt}=-[\langle I_x\rangle\langle 2S_y^2+4S_z^2\rangle+\langle 2I_y^2+4I_z^2\rangle\langle S_x\rangle$$

$$-2\langle I_yI_x\rangle\langle S_y\rangle-2\langle I_y\rangle\langle S_xS_y\rangle+2\langle I_z\rangle\langle S_xS_z\rangle$$
$$+2\langle I_zI_x\rangle\langle S_z\rangle]j(0)$$
$$-[\langle I_x\rangle\langle 2S_x^2+4S_y^2+4S_z^2\rangle+\langle 2I_x^2+4I_y^2+4I_z^2\rangle\langle S_x\rangle$$
$$-\langle I_y\rangle\langle S_xS_y+S_yS_x\rangle-\langle I_xI_y+I_yI_x\rangle\langle S_y\rangle$$
$$-\langle I_z\rangle\langle S_zS_x+S_xS_z\rangle-\langle I_zI_x+I_xI_z\rangle\langle S_z\rangle]j(\omega_0)$$
$$-[2\langle I_x\rangle\langle S_x^2+S_y^2\rangle+2\langle I_x^2+I_y^2\rangle\langle S_x\rangle$$
$$-2\langle I_zI_x+I_xI_z\rangle\langle S_z\rangle-2\langle I_z\rangle\langle S_zS_x+S_xS_z\rangle]j(2\omega_0). \qquad (4.45)$$

The results as they stand will not provide relaxation equations like (3.67) and (3.68) for I_r and S_r. It will clearly be necessary to assume that the components of **I** anticommute with one another, and that the components of **S** anticommute with one another. The well known situation in which these conditions are obeyed is that where the spins I and S are each equal to $\frac{1}{2}$. Then expressing the spin I in terms of Pauli operators by

$$I_x=\tfrac{1}{2}\sigma_x, \qquad I_y=\tfrac{1}{2}\sigma_y, \qquad I_z=\tfrac{1}{2}\sigma_z$$

and employing the relations (McConnell, 1960, p. 85)

$$\sigma_y\sigma_z + \sigma_z\sigma_y = 0, \qquad \sigma_y\sigma_z = i\sigma_z,$$

etc.,

$$\sigma_x^2 = \sigma_y^2 = \sigma_z^2 = E,$$

the identity operator, we deduce that

$$I_y I_z + I_z I_y = 0, \qquad I_y I_z = \tfrac{1}{2}iI_x,$$

etc.,

$$I_x^2 = I_y^2 = I_z^2 = \tfrac{1}{4}E, \qquad I(I+1) = \tfrac{3}{4}.$$

Using these and the corresponding equations for spin S we see that (4.44) and (4.45) reduce to

$$\frac{d\langle I_z + S_z \rangle}{dt} = -I(I+1)\{\tfrac{4}{3}j(\omega_0) + \tfrac{16}{3}j(2\omega_0)\}\langle I_z + S_z \rangle \qquad (4.46)$$

$$\frac{d\langle I_x + S_x \rangle}{dt} = -I(I+1)\{2j(0) + \tfrac{10}{3}j(\omega_0) + \tfrac{4}{3}j(2\omega_0)\}\langle I_x + S_x \rangle \qquad (4.47)$$

for two like nuclei of spin $\tfrac{1}{2}$. This result is exact.

We return to like nuclei of spin I and make the approximation $\hbar\omega_0 \ll kT$ proposed by Wangsness & Bloch (1953). Since the energy levels of the nucleus in the presence of the external field H_0 are the eigenvalues of $-\hbar\omega_0 I_z$, this approximation implies that the Zeeman splitting of the nuclear energy levels due to the external field is small compared with kT. We see from (1.16) that the approximation is a realistic one.

The ensemble averages $\langle I_r \rangle, \langle S_r \rangle$ which occur in expressions that we have been studying in this section arise from the perturbing Hamiltonian $\hbar G(t)$. They are therefore small quantities. We see from (4.44) and (4.45) that they are multiplied by quantities of the forms $\langle I_r I_{r'} \rangle, \langle S_r S_{r'} \rangle$, where r, $r' = x, y, z$. If now we expand the density matrix ρ as a series in powers of $\hbar\omega_0/kT$:

$$\rho = \rho_0 + \frac{\hbar\omega_0}{kT}\rho_1 + \cdots,$$

then in calculating to first order a product like

$$\langle I_r I_{r'} \rangle \langle S_x \rangle$$

we are entitled to employ the density operator for the unperturbed system when evaluating $\langle I_r I_{r'} \rangle$. For spin I the $2I+1$ states have equal probabilities. Hence the matrix representative of ρ_0 with respect to the basis constituted by the eigenfunctions of $\hbar\mathcal{H}_0$, that is of I_z, is given by

$$(\rho_0)_{mm'} = \frac{\delta_{mm'}}{2I+1}. \qquad (4.48)$$

In this approximation the ensemble average of A,

$$\langle A\rangle = \mathrm{tr}(\rho_0 A) = \sum_{mm'} (\rho_0)_{mm'} A_{m'm},$$

so that from (4.48)

$$\langle A\rangle = \frac{\mathrm{tr}\, A}{2I+1}. \tag{4.49}$$

Applying this equation to $A \equiv I_z^2$ we see that

$$\langle I_z^2\rangle = \frac{1}{2I+1} \sum_{m=-I}^{I} (I_z^2)_{mm}. \tag{4.50}$$

Since $(I_z^2)_{mm}$ is a diagonal matrix with the elements

$$I^2, (I-1)^2, \ldots, 1^2, 0, 1^2, \ldots, I^2$$

and since

$$1^2 + 2^2 + 3^2 + \cdots + I^2 = \frac{I(I+1)(2I+1)}{6},$$

it follows from (4.50) that

$$\langle I_z^2\rangle = \frac{I(I+1)}{3}.$$

Since by symmetry $\langle I_x^2\rangle = \langle I_y^2\rangle$ and since

$$\langle I_x^2 + I_y^2 + I_z^2\rangle = I(I+1),$$

we conclude that

$$\langle I_x^2\rangle = \langle I_y^2\rangle = \langle I_z^2\rangle = \frac{I(I+1)}{3} \tag{4.51}$$

in agreement with the exact result derived earlier for $I = \frac{1}{2}$.

In order to evaluate $\langle I_r I_{r'}\rangle$ for $r' \neq r$ we employ the explicit representations of I_x, I_y, I_z. The only non-vanishing elements are given by (Edmonds, 1968, p. 17)

$$(m|I_z|m) = m$$

$$(m\pm 1|I_x|m) = \tfrac{1}{2}[(j \mp m)(j \pm m + 1)]^{1/2}$$

$$(m\pm 1|I_y|m) = \mp\frac{i}{2}[(j \mp m)(j \pm m + 1)]^{1/2}.$$

On calculating $\mathrm{tr}(I_r I_{r'})$ we find without difficulty that its value is zero. Hence, from (4.49), $\langle I_r I_{r'}\rangle$ vanishes and so

$$\langle I_x I_y\rangle = \langle I_x I_z\rangle = \langle I_y I_x\rangle = \langle I_y I_z\rangle = \langle I_z I_x\rangle = \langle I_z I_y\rangle = 0. \tag{4.52}$$

On substituting (4.51), (4.52) and the corresponding equations for S_x, S_y, S_z into (4.44) and (4.45) we recover (4.46) and (4.47) as results that are good approximations for two like nuclei of spin I when $\hbar\omega_0 \ll kT$.

Equations (4.46) and (4.47) describe the relaxation processes when the system of two dipoles is influenced only by the perturbing Hamiltonian $\hbar G(t)$. The solutions obtained by equating to zero the time derivatives are

$$\langle I_x + S_x \rangle_0 = \langle I_z + S_z \rangle_0 = 0, \tag{4.53}$$

the zero subscript denoting steady state. If N is the number of two-dipole systems per unit volume and we assume that the interactions between the systems are negligible, the magnetization per unit volume

$$\mathbf{M} = N\gamma\hbar\langle \mathbf{I} + \mathbf{S} \rangle, \tag{4.54}$$

which shows that $\mathbf{I}$ and $\mathbf{S}$ have separately no physical significance. According to (4.53) and (4.54) the steady state value of M_z should be zero. However, as was discussed in section 1.1, the effect of the constant field in the z-direction is to give for the steady state

$$M_x = 0, \qquad M_z = M_0,$$

a finite constant. In order that M_z should relax to a non-vanishing $N\gamma\hbar\langle I_z + S_z \rangle_0$ under the combined action of the constant field H_0 and the perturbing field we replace $\langle I_z + S_z \rangle$ in (4.46) by $\langle I_z + S_z \rangle - \langle I_z + S_z \rangle_0$ and obtain

$$\frac{\mathrm{d}\langle I_z + S_z \rangle}{\mathrm{d}t} = -\tfrac{4}{3}\{j(\omega_0) + \tfrac{4}{3}j(2\omega_0)\}I(I+1)\{\langle I_z + S_z \rangle - \langle I_z + S_z \rangle_0\}. \tag{4.55}$$

We may then express (4.55) and (4.47) as

$$\frac{\mathrm{d}\langle I_z + S_z \rangle}{\mathrm{d}t} = -\frac{\langle I_z + S_z \rangle - \langle I_z + S_z \rangle_0}{T_1} \tag{4.56}$$

$$\frac{\mathrm{d}\langle I_x + S_x \rangle}{\mathrm{d}t} = -\frac{\langle I_x + S_x \rangle}{T_2}, \tag{4.57}$$

where

$$\frac{1}{T_1} = I(I+1)\{\tfrac{4}{3}j(\omega_0) + \tfrac{16}{3}j(2\omega_0)\} \tag{4.58}$$

$$\frac{1}{T_2} = I(I+1)\{2j(0) + \tfrac{10}{3}j(\omega_0) + \tfrac{4}{3}j(2\omega_0)\}. \tag{4.59}$$

The absence of $j(0)$ in (4.56) is a feature that will be found in all future expressions for the reciprocals of longitudinal relaxation times. It is possible to explain this feature by qualitative reasoning (Farrar & Becker, 1971, p. 49).

To complete the calculation of the relaxation times it is necessary to know the value of $j(\omega)$ for substitution into (4.58) and (4.59). This will depend on whether the dipolar interaction is intermolecular or intramolecular, and also on the particular model chosen to describe the thermal motion. It may be remarked that it is often found quite satisfactory to replace the arguments of the $j(\omega)$-functions by zero. When we do this, we say that we are working in the *extreme narrowing approximation*. In the present case (4.58) and (4.59) will then reduce to

$$\frac{1}{T_1} = \frac{1}{T_2} = \tfrac{20}{3}I(I+1)j(0), \tag{4.60}$$

and so the relaxation times will be equal.

4.3 Systems of unlike spins

We next examine nuclear magnetic relaxation in liquids undergoing thermal motion, when the relaxation is caused by dipolar interaction between non-identical particles. The interaction could be, for example, between two nuclei of unequal spin, or two nuclei of different elements with equal spin, or between a nucleus and an electron. We call the two spins I and S and their respective gyromagnetic ratios γ_I and γ_S. In the presence of a constant magnetic field H_0, which as previously is taken to be pointing in the positive z-direction of the laboratory system, there will be Larmor angular frequencies ω_I, ω_S given by

$$\omega_I = \gamma_I H_0, \qquad \omega_S = \gamma_S H_0. \tag{4.61}$$

The unperturbed Hamiltonian

$$\hbar \mathcal{H}_0 = -\gamma_I \hbar I_z H_0 - \gamma_S \hbar S_z H_0,$$

which from (4.61) is now expressible as

$$\hbar \mathcal{H}_0 = -\hbar(\omega_I I_z + \omega_S S_z). \tag{4.62}$$

The interaction Hamiltonian $\hbar G(t)$ is given by (4.17) and (4.18), and $j(\omega)$ is given by (4.20).

On account of the presence of the two angular frequencies in (4.62) the argument of the $j(\omega)$-function needs careful examination. Let us take the summation term

$$j(\omega_\alpha - \omega_\beta) \sum_{q=-2}^{2} (\alpha|A^{(q)}|\beta)(\beta'|A^{(q)+}|\alpha') \tag{4.63}$$

of (3.72). For $q=0$ we have from (4.27)

$$A^{(0)} = A^{(0)+} = \frac{2}{6^{1/2}} [2I_0 S_0 - \tfrac{1}{2}(I_+ S_- + I_- S_+)]. \tag{4.64}$$

Employing the notation of the last section we write

$$|\beta\rangle = |m_I\rangle|m_S\rangle$$

and deduce from (4.62) that

$$\omega_\beta = -(m_I\omega_I + m_S\omega_S). \qquad (4.65)$$

Let us also put

$$|\alpha\rangle = |m_I'\rangle|m_S'\rangle,$$

so that

$$\omega_\alpha = -(m_I'\omega_I + m_S'\omega_S). \qquad (4.66)$$

It will then follow, as in the previous section, that $(\alpha|I_0S_0|\beta)$ will be multiplied by $j(0)$ in (4.63).

Let us now consider the term $(\alpha|I_+S_-|\beta)$. As in (4.29) we put

$$I_+S_-|\beta\rangle = a_I b_S|m_I + 1\rangle|m_S - 1\rangle.$$

Hence for non-vanishing $(\alpha|I_+S_-|\beta)$ we must have

$$|\alpha\rangle = |m_I + 1\rangle|m_S - 1\rangle,$$

so that, from (4.66),

$$\omega_\alpha = -[(m_I + 1)\omega_I + (m_S - 1)\omega_S], \qquad (4.67)$$

and, from (4.65),

$$\omega_\alpha - \omega_\beta = \omega_S - \omega_I. \qquad (4.68)$$

We conclude that $(\alpha|I_+S_-|\beta)$ will be multiplied by $j(\omega_S - \omega_I)$ in the summation (4.63). Similarly $(\alpha|I_-S_+|\beta)$ will be multiplied by $j(\omega_I - \omega_S)$. Since $j(\omega)$ is an even function of ω, we may write the multiplying factor in both cases as $j(\omega_I - \omega_S)$.

Hence there is now a complication, not present for like spins, that we have two different values for $j(\omega_\alpha - \omega_\beta)$ in (4.63) when $q = 0$. However, this disadvantage can be offset by employing the relation (3.60), which tells us that

$$\omega_{\alpha'} - \omega_{\beta'} = \omega_\alpha - \omega_\beta.$$

Then, since

$$j(\omega_\alpha - \omega_\beta)A_{\alpha\beta}^{(0)}A_{\beta'\alpha'}^{(0)+} = j(\omega_{\beta'} - \omega_{\alpha'})A_{\beta'\alpha'}^{(0)+}A_{\alpha\beta}^{(0)},$$

we see that we obtain a non-vanishing contribution only when $(\alpha|I_+S_-|\beta)$ is multiplied by $(\beta'|I_-S_+|\alpha')$. This reduces the calculations considerably. Using the same approach to the other terms of $A^{(0)}$ and denoting generically by L each of the operators

$$\frac{4}{6^{1/2}}I_0S_0, \qquad -\frac{1}{6^{1/2}}I_+S_-, \qquad -\frac{1}{6^{1/2}}I_-S_+$$

we associate each $(\alpha|L|\beta)$ only with $(\beta'|L^+|\alpha')$. We conclude that

$$j(\omega_\alpha - \omega_\beta)(\alpha|A^{(0)}|\beta)(\beta'|A^{(0)}|\alpha')$$
$$= \tfrac{8}{3} j(0)(\alpha|I_0 S_0|\beta)(\beta'|I_0 S_0|\alpha')$$
$$+ \tfrac{1}{6} j(\omega_I - \omega_S)\{(\alpha|I_+ S_-|\beta)(\beta'|I_- S_+|\alpha') + (\alpha|I_- S_+|\beta)(\beta'|I_+ S_-|\alpha')\}.$$

$$(4.69)$$

For $q = \pm 1$ we have from (4.18)

$$A^{(-1)} = I_0 S_- + I_- S_0, \qquad A^{(1)} = -(I_0 S_+ + I_+ S_0).$$

From our above reasoning we know that the operators $I_0 S_-$ and $I_0 S_+$ will give to (4.63) a contribution proportional to $j(\omega_S)$ and that the operators $I_- S_0$ and $I_+ S_0$ will give a contribution proportional to $j(\omega_I)$. Moreover, if L' denotes any one of the operators $I_0 S_-, I_- S_0, I_0 S_+, I_+ S_0$, the only non-vanishing products of matrix elements that occur are those of the form $(\alpha|L'|\beta)(\beta'|L'^+|\alpha')$. We therefore obtain

$$j(\omega_\alpha - \omega_\beta)\{(\alpha|A^{(-1)}|\beta)(\beta'|A^{(-1)+}|\alpha') + (\alpha|A^{(1)}|\beta)(\beta'|A^{(1)+}|\alpha')\}$$
$$= j(\omega_I)\{(\alpha|I_- S_0|\beta)(\beta'|I_+ S_0|\alpha') + (\alpha|I_+ S_0|\beta)(\beta'|I_- S_0|\alpha')\}$$
$$+ j(\omega_S)\{(\alpha|I_0 S_-|\beta)(\beta'|I_0 S_+|\alpha') + (\alpha|I_0 S_+|\beta)(\beta'|I_0 S_-|\alpha')\}. (4.70)$$

For $q = \pm 2$ we have from (4.18)

$$A^{(-2)} = I_- S_-, \qquad A^{(2)} = I_+ S_+.$$

We then deduce that

$$j(\omega_\alpha - \omega_\beta)\{(\alpha|A^{(-2)}|\beta)(\beta'|A^{(-2)+}|\alpha') + (\alpha|A^{(2)}|\beta)(\beta'|A^{(2)+}|\alpha')\}$$
$$= j(\omega_I + \omega_S)\{(\alpha|I_- S_-|\beta)(\beta'|I_+ S_+|\alpha') + (\alpha|I_+ S_+|\beta)(\beta'|I_- S_-|\alpha')\}.$$

$$(4.71)$$

Returning to (3.72) we see that on account of (3.60) the second term on the right hand side of (3.72) is equal to the first one. Let us therefore discuss

$$\delta_{\alpha\beta} j(\omega_\sigma - \omega_{\beta'})(\sigma|A^{(q)}|\alpha')(\beta'|A^{(q)+}|\sigma). \qquad (4.72)$$

The $I_0 S_0$-part of $A^{(0)+}$ will provide a non-vanishing contribution to $(\beta'|A^{(0)+}|\sigma)$ only when $\omega_{\beta'} - \omega_\sigma$ vanishes. On account of (3.60) and the Kronecker delta in (4.72) it will follow that

$$\omega_\sigma - \omega_{\alpha'} = \omega_\sigma + \omega_\beta - \omega_{\beta'} - \omega_\alpha = \omega_\sigma - \omega_{\beta'} = 0,$$

and so the only non-vanishing contribution to $(\sigma|A^{(0)}|\alpha')$ will come from the $I_0 S_0$-part of $A^{(0)}$. Similarly we see from (4.68) that the only non-vanishing contribution of the $I_+ S_-$-part of $A^{(0)+}$ to $(\beta'|A^{(0)+}|\sigma)$ comes from

$$\omega_{\beta'} - \omega_\sigma = \omega_S - \omega_I.$$

Then

$$\omega_\sigma - \omega_{\alpha'} = \omega_\sigma - \omega_{\beta'} + \omega_\beta - \omega_\alpha = \omega_\sigma - \omega_{\beta'} = \omega_I - \omega_S$$

and the only non-vanishing contribution to $(\sigma|A^{(0)}|\alpha')$ will come from the I_-S_+-part of $A^{(0)}$. A similar argument will hold for the $q = \pm 1, \pm 2$ terms of (4.72) and for

$$\delta_{\alpha'\beta'}j(\omega_\sigma - \omega_\beta)(\sigma|A^{(q)}|\beta)(\alpha|A^{(q)+}|\sigma).$$

We may therefore use the pattern provided by (4.69)–(4.71) for (4.63) to write out in full the terms within the braces of (3.72).

We now apply the method of the previous section to calculate relaxation times from (3.72).

For $q = 0$ we deduce from (4.69) and (3.72) that the contribution to $d\langle I_r\rangle/dt$,

$$\sum_{\substack{\alpha\alpha'\beta\beta'}} \rho(t)_{\beta\beta'} \left\{ \tfrac{16}{3}j(0)(\alpha|I_0S_0|\beta)(\beta'|I_0S_0|\alpha') \right.$$

$$+ \tfrac{1}{3}j(\omega_I - \omega_S)[(\alpha|I_+S_-|\beta)(\beta'|I_-S_+|\alpha') + (\alpha|I_-S_+|\beta)(\beta'|I_+S_-|\alpha')]$$

$$- \delta_{\alpha\beta}\left[\sum_\sigma \{\tfrac{8}{3}j(0)(\sigma|I_0S_0|\alpha')(\beta'|I_0S_0|\sigma)\right.$$

$$\left. + \tfrac{1}{6}j(\omega_I - \omega_S)[(\sigma|I_+S_-|\alpha')(\beta'|I_-S_+|\sigma) + (\sigma|I_-S_+|\alpha')(\beta'|I_+S_-|\sigma)]\}\right]$$

$$- \delta_{\alpha'\beta'}\left[\sum_\sigma \{\tfrac{8}{3}j(0)(\sigma|I_0S_0|\beta)(\alpha|I_0S_0|\sigma)\right.$$

$$\left.\left. + \tfrac{1}{6}j(\omega_I - \omega_S)[(\sigma|I_+S_-|\beta)(\alpha|I_-S_+|\sigma) + (\sigma|I_-S_+|\beta)(\alpha|I_+S_-|\sigma)]\}\right]\right\}$$

$$\times (\alpha'|I_r|\alpha)$$

$$= \langle\{\tfrac{8}{3}j(0)(2I_0S_0I_rI_0S_0 - I_0S_0I_0S_0I_r - I_rI_0S_0I_0S_0)$$

$$+ \tfrac{1}{6}j(\omega_I - \omega_S)(2I_-S_+I_rI_+S_- - I_-S_+I_+S_-I_r - I_rI_-S_+I_+S_-$$

$$+ 2I_+S_-I_rI_-S_+ - I_+S_-I_-S_+I_r - I_rI_+S_-I_-S_+)\}\rangle$$

$$= \tfrac{8}{3}j(0)\langle[[I_0S_0, I_r], I_0S_0]\rangle$$

$$+ \tfrac{1}{6}j(\omega_I - \omega_S)\langle[[I_+S_-, I_r], I_-S_+] + \text{adj}\rangle. \tag{4.73}$$

For $q = \pm 1$ the contribution from (4.70) to $d\langle I_r\rangle/dt$,

$$\sum_{\substack{\alpha\alpha'\beta\beta'}} \rho(t)_{\beta\beta'} \left\{ j(\omega_I)[2(\alpha|I_-S_0|\beta)(\beta'|I_+S_0|\alpha') + 2(\alpha|I_+S_0|\beta)(\beta'|I_-S_0|\alpha') \right.$$

$$- \delta_{\alpha\beta}\sum_\sigma \{(\sigma|I_-S_0|\alpha')(\beta'|I_+S_0|\sigma) + (\sigma|I_+S_0|\alpha')(\beta'|I_-S_0|\sigma)\}$$

$$- \delta_{\alpha'\beta'}\sum_\sigma \{(\sigma|I_-S_0|\beta)(\alpha|I_+S_0|\sigma) + (\sigma|I_+S_0|\beta)(\alpha|I_-S_0|\sigma)\}]$$

$$+ j(\omega_S)[2(\alpha|I_0S_-|\beta)(\beta'|I_0S_+|\alpha') + 2(\alpha|I_0S_+|\beta)(\beta'|I_0S_-|\alpha')$$

$$-\delta_{\alpha\beta}\sum_{\sigma}\{(\sigma|I_0S_-|\alpha')(\beta'|I_0S_+|\sigma)+(\sigma|I_0S_+|\alpha')(\beta'|I_0S_-|\sigma)\}$$

$$-\delta_{\alpha'\beta'}\sum_{\sigma}\{(\sigma|I_0S_-|\beta)(\alpha'|I_0S_+|\sigma)+(\sigma|I_0S_+|\beta)(\alpha|I_0S_-|\sigma)\}\Big]\Big\}\times(\alpha'|I_r|\alpha)$$

$$=j(\omega_I)\langle[[I_+S_0,I_r],I_-S_0]+\mathrm{adj}\rangle$$

$$+j(\omega_S)\langle[[I_0S_+,I_r],I_0S_-]+\mathrm{adj}\rangle. \tag{4.74}$$

The contribution from (4.71) for $q=\pm2$,

$$\sum_{\alpha\alpha'\beta\beta'}\rho(t)_{\beta\beta'}\,j(\omega_I+\omega_S)\Big\{2(\alpha|I_-S_-|\beta)(\beta'|I_+S_+|\alpha')+2(\alpha|I_+S_+|\beta)(\beta'|I_-S_-|\alpha')$$

$$-\delta_{\alpha\beta}\sum_{\sigma}[(\sigma|I_-S_-|\alpha')(\beta'|I_+S_+|\sigma)+(\sigma|I_+S_+|\alpha')(\beta'|I_-S_-|\sigma)]$$

$$-\delta_{\alpha'\beta'}\sum_{\sigma}[(\sigma|I_-S_-|\beta)(\alpha|I_+S_+|\sigma)+(\sigma|I_+S_+|\beta)(\alpha|I_-S_-|\sigma)]\Big\}$$

$$\times(\alpha'|I_r|\alpha)$$

$$=j(\omega_I+\omega_S)\langle[[I_+S_+,I_r],I_-S_-]+\mathrm{adj}\rangle. \tag{4.75}$$

The evaluation of the expressions in (4.73)–(4.75) presents no great difficulty. The double commutators are simplified by applying (4.36) and (4.37), and noting (4.33). We continue to use the approximations (4.51) and (4.52), and we replace $\langle I_z\rangle$ by $\langle I_z\rangle-\langle I_z\rangle_0$ for the reason explained at the end of the previous section. It is then deduced that

$$\frac{d(\langle I_z\rangle-\langle I_z\rangle_0)}{dt}=-\tfrac{4}{9}j(\omega_I-\omega_S)S(S+1)(\langle I_z\rangle-\langle I_z\rangle_0)$$

$$+\tfrac{4}{9}j(\omega_I-\omega_S)I(I+1)(\langle S_z\rangle-\langle S_z\rangle_0)$$

$$-\tfrac{4}{3}S(S+1)j(\omega_I)(\langle I_z\rangle-\langle I_z\rangle_0)-\tfrac{8}{3}I(I+1)j(\omega_I+\omega_S)(\langle S_z\rangle-\langle S_z\rangle_0)$$

$$-\tfrac{8}{3}S(S+1)j(\omega_I+\omega_S)(\langle I_z\rangle-\langle I_z\rangle_0), \tag{4.76}$$

$$\frac{d\langle I_x\rangle}{dt}=-\tfrac{8}{9}j(0)S(S+1)\langle I_x\rangle-\tfrac{2}{9}j(\omega_I-\omega_S)S(S+1)\langle I_x\rangle-\tfrac{2}{3}j(\omega_I)(S+1)\langle I_x\rangle$$

$$-\tfrac{4}{3}j(\omega_S)S(S+1)\langle I_x\rangle-\tfrac{4}{3}j(\omega_I+\omega_S)S(S+1)\langle I_x\rangle. \tag{4.77}$$

We see that the equations for $\langle I_z\rangle-\langle I_z\rangle_0$ and $\langle S_z\rangle-\langle S_z\rangle_0$ are coupled, but that there is no such coupling between $\langle I_x\rangle$ and $\langle S_x\rangle$.

Equations (4.76), (4.77) and those obtained by interchanging I and S may be expressed as follows:

$$\frac{d(\langle I_z\rangle-\langle I_z\rangle_0)}{dt}=-\frac{\langle I_z\rangle-\langle I_z\rangle_0}{T_1^{II}}-\frac{\langle S_z\rangle-\langle S_z\rangle_0}{T_1^{IS}}, \tag{4.78}$$

$$\frac{d(\langle S_z\rangle-\langle S_z\rangle_0)}{dt}=-\frac{\langle I_z\rangle-\langle I_z\rangle_0}{T_1^{SI}}-\frac{\langle S_z\rangle-\langle S_z\rangle_0}{T_1^{SS}}, \tag{4.79}$$

$$\frac{\mathrm{d}\langle I_x \rangle}{\mathrm{d}t} = -\frac{\langle I_x \rangle}{T_2^I}, \qquad \frac{\mathrm{d}\langle S_x \rangle}{\mathrm{d}t} = -\frac{\langle S_x \rangle}{T_2^S}, \tag{4.80}$$

where

$$\frac{1}{T_1^{II}} = S(S+1)\{\tfrac{4}{9}j(\omega_I - \omega_S) + \tfrac{4}{3}j(\omega_I) + \tfrac{8}{3}j(\omega_I + \omega_S)\} \tag{4.81}$$

$$\frac{1}{T_1^{SS}} = I(I+1)\{\tfrac{4}{9}j(\omega_I - \omega_S) + \tfrac{4}{3}j(\omega_S) + \tfrac{8}{3}j(\omega_I + \omega_S)\} \tag{4.82}$$

$$\frac{1}{T_1^{IS}} = I(I+1)\{ -\tfrac{4}{9}j(\omega_I - \omega_S) + \tfrac{8}{3}j(\omega_I + \omega_S)\} \tag{4.83}$$

$$\frac{1}{T_1^{SI}} = S(S+1)\{ -\tfrac{4}{9}j(\omega_I - \omega_S) + \tfrac{8}{3}j(\omega_I + \omega_S)\} \tag{4.84}$$

$$\frac{1}{T_2^I} = S(S+1)\{\tfrac{8}{9}j(0) + \tfrac{2}{9}j(\omega_I - \omega_S) + \tfrac{4}{3}j(\omega_I) + \tfrac{2}{3}j(\omega_S) + \tfrac{4}{3}j(\omega_I + \omega_S)\} \tag{4.85}$$

$$\frac{1}{T_2^S} = I(I+1)\{\tfrac{8}{9}j(0) + \tfrac{2}{9}j(\omega_I - \omega_S) + \tfrac{4}{3}j(\omega_I) + \tfrac{2}{3}j(\omega_S) + \tfrac{4}{3}j(\omega_I + \omega_S)\}. \tag{4.86}$$

In his discussion of relaxation by dipolar coupling, Abragam (1961, pp. 278, 289–96) expressed his results in terms of the functions $J^{(0)}(\omega)$, $J^{(1)}(\omega)$, $J^{(2)}(\omega)$ defined by

$$J^{(l)}(\omega) = \int_{-\infty}^{\infty} \langle F^{(l)}(0) F^{(l)}(\tau)^* \rangle \, \mathrm{e}^{-i\omega\tau} \, \mathrm{d}\tau, \tag{4.87}$$

where from (4.8)

$$F^{(0)}(t) = \frac{1 - 3\cos^2 \theta(t)}{r(t)^3} = -\left(\frac{16\pi}{5}\right)^{1/2} \frac{Y_{20}(\theta(t), \phi(t))}{r(t)^3}$$

$$F^{(1)}(t) = \frac{\sin \theta(t) \cos \theta(t) \, \mathrm{e}^{-i\phi(t)}}{r(t)^3} = \left(\frac{8\pi}{15}\right)^{1/2} \frac{Y_{2,-1}(\theta(t), \phi(t))}{r(t)^3} \tag{4.88}$$

$$F^{(2)}(t) = \frac{\sin^2 \theta(t) \, \mathrm{e}^{-2i\phi(t)}}{r(t)^3} = \left(\frac{32\pi}{15}\right)^{1/2} \frac{Y_{2,-2}(\theta(t), \phi(t))}{r(t)^3}.$$

On comparing (4.88) with (4.19) we see that

$$(\tfrac{3}{8})^{1/2}\gamma_I\gamma_S\hbar F^{(0)}(t) = H_0(t)$$
$$-\tfrac{3}{2}\gamma_I\gamma_S\hbar F^{(1)}(t) = H_{-1}(t)$$
$$-\tfrac{3}{4}\gamma_I\gamma_S\hbar F^{(2)}(t) = H_{-2}(t).$$

Hence

$$\gamma_I^2\gamma_S^2\hbar^2\langle F^{(0)}(0)F^{(0)}(\tau)^*\rangle = \tfrac{8}{3}\langle H_0(0)H_0(\tau)\rangle$$

$$\gamma_I^2\gamma_S^2\hbar^2\langle F^{(1)}(0)F^{(1)}(\tau)^*\rangle = \tfrac{4}{9}\langle H_{-1}(0)H_{-1}(\tau)^*\rangle = -\tfrac{4}{9}\langle H_{-1}(0)H_1(\tau)\rangle$$

$$= \tfrac{4}{9}\langle H_0(0)H_0(\tau)\rangle,$$

by (3.41), and similarly

$$\gamma_I^2\gamma_S^2\hbar^2\langle F^{(2)}(0)F^{(2)}(\tau)^*\rangle = \tfrac{16}{9}\langle H_0(0)H_0(\tau)\rangle.$$

Thus, from (3.48) and (4.87),

$$j(\omega) = \tfrac{3}{16}\gamma_I^2\gamma_S^2\hbar^2 J^{(0)}(\omega) = \tfrac{9}{8}\gamma_I^2\gamma_S^2\hbar^2 J^{(1)}(\omega) = \tfrac{9}{32}\gamma_I^2\gamma_S^2\hbar^2 J^{(2)}(\omega). \quad (4.89)$$

On substituting from (4.89) into Abragam's expressions for relaxation times we find complete agreement with (4.58) and (4.59) for like spins, and with (4.81)–(4.86) for unlike spins. The advantage of our presentation is that it shows that the values of the relaxation times depend on the calculation of the single function $j(\omega)$.

In the extreme narrowing approximation defined in the last section, (4.81)–(4.86) reduce to

$$\frac{1}{T_1^{II}} = \tfrac{40}{9}S(S+1)j(0), \qquad \frac{1}{T_1^{SS}} = \tfrac{40}{9}I(I+1)j(0)$$

$$\frac{1}{T_1^{IS}} = \tfrac{20}{9}I(I+1)j(0), \qquad \frac{1}{T_1^{SI}} = \tfrac{20}{9}S(S+1)j(0) \qquad (4.90)$$

$$\frac{1}{T_2^{I}} = \tfrac{40}{9}S(S+1)(0), \qquad \frac{1}{T_2^{S}} = \tfrac{40}{9}I(I+1)j(0).$$

We compare these equations with (4.60) for like spins:

$$\frac{1}{T_1} = \frac{1}{T_2} = \tfrac{20}{3}I(I+1)j(0). \qquad (4.91)$$

In the case of $S=I$, $\gamma_S=\gamma_I$, (4.90) do not reduce to (4.91). Indeed the interdependence of the equations for the longitudinal relaxation still persists. Moreover

$$\frac{1/T_2}{1/T_2^I} = \frac{3}{2},$$

which shows that for transverse relaxation the coupling between like spins is stronger than that between unlike spins.

5

Relaxation by intermolecular dipolar interactions

5.1 Intermolecular interactions

It was seen in section 4.1 that dipole–dipole interaction depends on both the distance between the two dipoles and on the orientation of the line joining the dipoles. If the dipoles belong to nuclei in the same molecule, and if we regard the molecule as rigid, the time dependence of the interaction is due entirely to the rotation of the molecule. When the nuclei are in different molecules, the time dependence will in general arise from both the translational and rotational motion of the molecules.

We now consider the nuclear magnetic relaxation that results from random translational motion of the molecules. The investigations will be performed firstly by employing random walk models and later by Brownian motion models.

5.2 Random walk models

Torrey (1953) applied his theory of random walk, described in subsection 2.4.1, to investigate nuclear spin relaxation by translational random motion. He assumed that the spins are at the centres of identical hard spherical molecules. Then the rotational motion of the individual molecules did not enter into this calculations, since it did not affect the length or the orientation of the line joining the two spins. He devoted his attention chiefly to *isotropic diffusion*, which he defined by the two conditions:

(*a*) the $A(\boldsymbol{\rho})$ of (2.27) depends only on the modulus of $\boldsymbol{\rho}$, so that we write $A(\boldsymbol{\rho})$ as $A(\rho)$,

(*b*) the density of the spins in uniform; we take it to be N spins per unit volume.

Let us evaluate $j(\omega)$ for isotropic diffusion from (4.20). In order to calculate the ensemble average we multiply

$$r(0)^{-3}Y_{20}(\theta(0), \phi(0))r(t)^{-3}Y_{20}(\theta(t), \phi(t))$$

by the joint probability density function and integrate with respect to $\mathbf{r}(0)$ and $\mathbf{r}(t)$. Employing the notation and results of Chapter 2 we replace the joint probability density function by

$$w(\mathbf{r}_j(0) - \mathbf{r}_i(0), \mathbf{r}(0); \mathbf{r}_j(t) - \mathbf{r}_i(t), \mathbf{r}(t))p(\mathbf{r}_j(0) - \mathbf{r}_i(0), \mathbf{r}(0)). \qquad (5.1)$$

When we integrate $p(\mathbf{r}_j(0) - \mathbf{r}_i(0), \mathbf{r}(0)) \, \mathrm{d}^3\mathbf{r}(0)$ over a unit volume which is large compared with the dimensions of the dipole–dipole systems, we obtain N. Hence we may replace p by N in (5.1). Moreover we have from (2.37) that the conditional probability density w in (5.1) is equal to

$$\frac{1}{8\pi^3} \int \exp\left\{ -i(\boldsymbol{\rho}\cdot\mathbf{r}(t) - \mathbf{r}(0)) - \frac{2t}{\tau}(1 - A(\rho)) \right\} \mathrm{d}^3\boldsymbol{\rho}, \qquad (5.2)$$

where τ is the mean time between flights. Collecting our results and putting $\gamma_I = \gamma_S = \gamma$ in (4.20), since the molecules are identical, we find that

$$j(\omega) = \frac{3N\gamma^4\hbar^2}{40\pi^2} \int_{-\infty}^{\infty} \mathrm{d}t\, e^{-i\omega t} \iint \mathrm{d}^3\mathbf{r}(0)\, \mathrm{d}^3\mathbf{r}(t) r(0)^{-3} Y_{20}(\theta(0), \phi(0))$$

$$\times r(t)^{-3} Y_{20}(\theta(t), \phi(t)) \int_0^{\infty} \exp\left\{ -i(\boldsymbol{\rho}\cdot\mathbf{r}(t) - \mathbf{r}(0)) - \frac{2t}{\tau}(1 - A(\rho)) \right\} \mathrm{d}^3\boldsymbol{\rho}, \quad (5.3)$$

where the limits of integration of $r(0)$ and $r(t)$ are from the minimum distance of approach of the spins d_0 to infinity.

To proceed with the integrations we denote by θ', ϕ' the polar angles of $\boldsymbol{\rho}$ and put (Edmonds, 1968, p. 81)

$$e^{i(\boldsymbol{\rho}\cdot\mathbf{r}(0))} = 4\pi \left(\frac{\pi}{2\rho r(0)}\right)^{1/2} \sum_{l=0}^{\infty} \sum_{n=-l}^{l} i^l Y_{ln}^*(\theta(0), \phi(0)) Y_{ln}(\theta', \phi') J_{l+1/2}(\rho r(0))$$

$$(5.4)$$

$$e^{-i(\boldsymbol{\rho}\cdot\mathbf{r}(t))} = 4\pi \left(\frac{\pi}{2\rho r(t)}\right)^{1/2} \sum_{l'=0}^{\infty} \sum_{n'=-l'}^{l'} (-i)^{l'} Y_{l'n'}^*(\theta(t), \phi(t)) Y_{l'n'}(\theta', \phi') J_{l'+1/2}(\rho r(t)),$$

where J_m is a Bessel function of order m. Expressing $\mathrm{d}^3\mathbf{r}(0)$, $\mathrm{d}^3\mathbf{r}(t)$, $\mathrm{d}^3\boldsymbol{\rho}$ by

$$\mathrm{d}^3\mathbf{r}(0) = r(0)^2 \, \mathrm{d}r(0) \sin\theta(0)\, \mathrm{d}\theta(0)\, \mathrm{d}\phi(0)$$

$$\mathrm{d}^3\mathbf{r}(t) = r(t)^2 \, \mathrm{d}r(t) \sin\theta(t)\, \mathrm{d}\theta(t)\, \mathrm{d}\phi(t)$$

$$\mathrm{d}^3\boldsymbol{\rho} = \rho^2 \, \mathrm{d}\rho \sin\theta' \, \mathrm{d}\theta' \, \mathrm{d}\phi'$$

and employing the relation (B.5) of Appendix B

$$\iint Y_{jm}^*(\theta, \phi) Y_{j'm'}(\theta, \phi) \sin\theta\, \mathrm{d}\theta\, \mathrm{d}\phi = \delta_{jj'}\delta_{mm'}$$

we deduce from (5.3) and (5.4) that

$$l = l' = 2, \qquad n = n' = 0$$

$$j(\omega) = \frac{3N\gamma^4\hbar^2}{5} \int_{-\infty}^{\infty} \mathrm{d}t\, e^{-i\omega t} \int_0^{\infty} \mathrm{d}\rho\, \rho \exp\left[-\frac{2t}{\tau}(1 - A(\rho)) \right] \left(\int_{d_0}^{\infty} r^{-3/2} J_{5/2}(\rho r)\, \mathrm{d}r \right)^2,$$

$$(5.5)$$

where r stands for either $r(0)$ or $r(t)$. In order to calculate the last integral we note the property of Bessel functions (McLachlan, 1934, p. 25)

$$\frac{\mathrm{d}}{\mathrm{d}z}(z^{-m}J_m(z)) = -z^{-m}J_{m+1}(z),$$

which yields

$$\int_b^\infty z^{-m}J_{m+1}(z)\,\mathrm{d}z = b^{-m}J_m(b).$$

On making the replacements

$$z\mapsto \rho r, \qquad b\mapsto \rho d_0, \qquad m\mapsto \tfrac{3}{2}$$

we obtain

$$\int_{d_0}^\infty J_{5/2}(\rho r)r^{-3/2}\,\mathrm{d}r = \rho^{-1}d_0^{-3/2}J_{3/2}(\rho d_0),$$

and so (5.5) becomes

$$j(\omega) = \frac{3N\gamma^4\hbar^2}{5d_0^3}\int_{-\infty}^\infty \mathrm{d}t\, e^{-i\omega t}\int_0^\infty \mathrm{d}\rho \exp\left[-\frac{2t}{\tau}(1-A(\rho))\right]\rho^{-1}(J_{3/2}(\rho d_0))^2.$$

(5.6)

On comparing (5.6) with (5.3) we see that the integral over ρ is a real multiple of the autocorrelation function of $Y_{20}(\theta(t),\phi(t))r(t)^{-3}$, which is itself real. Then from our discussions of spectral densities in section 2.3 and in particular from (2.23) we may argue that (5.6) is expressible as

$$j(\omega) = \frac{6N\gamma^4\hbar^2}{5d_0^3}\,\mathrm{Re}\int_0^\infty \mathrm{d}t\, e^{-i\omega t}\int_0^\infty \mathrm{d}\rho \exp\left[-\frac{2t}{\tau}(1-A(\rho))\right]\rho^{-1}(J_{3/2}(\rho d_0))^2.$$

Permuting the order of integration and noting that

$$\mathrm{Re}\int_0^\infty \mathrm{d}t\,\exp\left\{-t\left[i\omega + \frac{2}{\tau}(1-A(\rho))\right]\right\} = \frac{\tfrac{1}{2}\tau(1-A(\rho))}{(1-A(\rho))^2 + \tfrac{1}{4}\omega^2\tau^2}$$

we conclude that

$$j(\omega) = \frac{3\pi N\gamma^4\hbar^2\tau}{5d_0^3}\int_0^\infty \frac{\mathrm{d}\rho}{\rho}\,\frac{1-A(\rho)}{[1-A(\rho)]^2 + \tfrac{1}{4}\omega^2\tau^2}\,(J_{3/2}(\rho d_0))^2. \qquad (5.7)$$

The longitudinal and transverse relaxation times may then be found from (4.58), (4.59) and (5.7), when the value of $A(\rho)$ is prescribed. Whatever its value, by taking absolute values of both sides of (2.27) and remembering that $P_1(\mathbf{r})$ is real and positive we see that

$$|A(\rho)| \leqslant \int P_1(\mathbf{r})\left|e^{i(\boldsymbol{\rho}\cdot\mathbf{r})}\right|\mathrm{d}^3\mathbf{r}$$

$$= \int P_1(\mathbf{r})\,\mathrm{d}^3\mathbf{r},$$

and therefore that

$$|A(\rho)| \leqslant 1. \tag{5.8}$$

Using only the inequality (5.8) Torrey expresses $j(\omega)$ for certain approximations in terms of integrals $\mathscr{I}_1$ and $\mathscr{I}_2$ defined by

$$\mathscr{I}_1 = \int_0^\infty \rho^{-1}(1-A(\rho))[J_{3/2}(\rho d_0)]^2$$

$$\mathscr{I}_2 = \int_0^\infty \rho^{-1}(1-A(\rho))^{-1}[J_{3/2}(\rho d_0)]^2.$$

Thus for $\omega\tau \gg 1$

$$j(\omega) = \frac{12\pi N\gamma^4\hbar^2\mathscr{I}_1}{5d_0^3\omega^2\tau},$$

and consequently from (4.58)

$$\frac{1}{T_1} = \frac{32\pi N\gamma^4\hbar^2 I(I+1)\mathscr{I}_1}{5d_0^2\omega_0^2\tau} \qquad (\omega_0\tau \gg 1),$$

so that T_1 is proportional to the mean lifetime between flights τ, if τ is large compared with the Larmor period $2\pi/\omega_0$. If on the other hand $\omega_0\tau \ll 1$, it is found that

$$\frac{1}{T_1} = \frac{4\pi N\gamma^4\hbar^2\tau I(I+1)\mathscr{I}_2}{d_0^3} \qquad (\omega_0\tau \ll 1),$$

so that T_1 is proportional to τ^{-1}.

Another case considered by Torrey in which $A(\rho)$ is not specified is that where the mean squared flight distance $\langle r^2 \rangle \gg d_0^2$, the square of the minimum distance of approach of the dipoles. Torrey argues that this condition ensures that the value of $P_1(\mathbf{r})$ becomes very small and that consequently, from (2.27), $1-A(\rho)$ may be replaced by unity in the integrand of (5.7). Doing this and employing the relation (Magnus & Oberhettinger, 1949, p. 35)

$$\int_0^\infty \rho^{-1}(J_{3/2}(\rho d_0))^2\,\mathrm{d}\rho = \tfrac{1}{3}$$

one obtains

$$j(\omega) = \frac{\pi N\gamma^4\hbar^2\tau}{5d_0^3}\frac{1}{1+\tfrac{1}{4}\omega^2\tau^2}.$$

Then from (4.58)

$$\frac{1}{T_1} = \frac{4\pi N\gamma^4\hbar^2\tau I(I+1)}{15d_0^3}\left(\frac{1}{1+\tfrac{1}{4}\omega_0^2\tau^2} + \frac{4}{1+\omega_0^2\tau^2}\right). \tag{5.9}$$

The last term in the bracket differs from that of Torrey's equation (36) by a factor of two. This is due to his use of the incorrect equation (34) of Bloembergen *et al.* (1948), which should read as in (1.56).

Torrey also made an analytical study of expressions for relaxation times resulting from the special forms of $A(\rho)$ given by

$$A(\rho)=\frac{1}{1+D\tau\rho^2} \tag{5.10}$$

$$A(\rho)=\frac{\sin l\rho}{l\rho}. \tag{5.11}$$

Equation (5.11) corresponds to jumps of equal length l in random directions. A general result of these calculations of Torrey and of similar calculations is that in the extreme narrowing approximation the value of T_1^{-1} does not depend very much on the chosen form of $A(\rho)$ (Hertz, 1967, p. 177). We shall later consider further implications of the Torrey theory by regarding it as a special case of another theory that will now be described.

Harmon & Muller (1969) generalized the Torrey theory by relaxing condition (*b*), for isotropic diffusion given at the beginning of this section, while retaining condition (*a*). This they did by replacing the spin density N per unit volume by a spherically symmetric $Ng(r(0))$, where $\int g(r(0))\,d^3r(0)=1$. Equation (5.3) now becomes

$$j(\omega)=\frac{3N\gamma^4\hbar^2}{40\pi^2}\int_{-\infty}^{\infty}dt\,e^{-i\omega t}\iint d^3r(0)\,d^3r(t)g(r(0))r(0)^{-3}Y_{20}(\theta(0),\phi(0)) \tag{5.12}$$

$$\times r(t)^{-3}Y_{20}(\theta(t),\phi(t))\int_0^{\infty}\exp\left\{-i(\rho\cdot r(t)-r(0))-\frac{2t}{\tau}(1-A(\rho))\right\}d^3\rho.$$

It is assumed that there exists a length $b>d_0$ such that $g(r(0))$ has the following properties:

(α) $g(r(0))=0$ for $0<r(0)<d_0$;
(β) $g(r(0))$ exhibits damped oscillations about unity for $d_0<r(0)<b$;
(γ) $g(r(0))\approx 1$ for $r(0)\geqslant b$.

If $b=d_0$, we have $g(r(0))$ identical with the hard sphere distribution function studied by Torrey. On employing the above three properties of $g(r(0))$ and the mathematical methods used in the derivation of (5.7), (5.12) is expressed as

$$j(\omega)=\frac{3\pi\gamma^4\hbar^2N\tau}{5d_0^{3/2}b^{3/2}}\int_0^{\infty}\left\{\frac{1-A(\rho)}{[1-A(\rho)]^2+(\tfrac{1}{2}\omega\tau)^2}\right\}J_{3/2}(\rho d_0)J_{3/2}(\rho b)\rho^{-1}\,d\rho$$

$$+\frac{3\pi\gamma^4\hbar^2 N\tau}{5d_0^{3/2}}\int_0^\infty \left\{\frac{1-A(\rho)}{[1-A(\rho)]^2+(\tfrac{1}{2}\omega\tau)^2}\right\}J_{3/2}(\rho d_0)\,\mathrm{d}\rho$$

$$\times\int_{d_0}^b J_{5/2}(\rho r(0))g(r(0))r(0)^{-3/2}\,\mathrm{d}r(0). \tag{5.13}$$

According to (2.41), (2.43) and (2.44) the value of $A(\rho)$ given by

$$A(\rho)=1-D'\tau\rho^2+O(\rho^4) \tag{5.14}$$

leads in Torrey's theory to diffusive motion of the molecules that carry the spins. We examine what (5.13) becomes when $A(\rho)$ has this value and we work in the approximation $\omega\tau\ll1$. Then (5.13) reduces to

$$j(0)=\frac{3\pi\gamma^4\hbar^2 N}{5d_0^{3/2}b^{3/2}D'}\int_0^\infty J_{3/2}(\rho d_0)J_{3/2}(\rho b)\rho^{-3}\,\mathrm{d}\rho$$

$$+\frac{3\pi\gamma^4\hbar^2 N}{5d_0^{3/2}D'}\int_{d_0}^b g(r(0))r(0)^{-3/2}\,\mathrm{d}r(0)\int_0^\infty J_{3/2}(\rho d_0)J_{5/2}(\rho r(0))\rho^{-2}\,\mathrm{d}\rho,$$

where we have interchanged the order of integration in the double integral. On expressing the integrals as hypergeometric functions (Magnus & Oberhettinger, 1949, pp. 7, 10, 35) we obtain

$$j(0)=\frac{\pi\gamma^4\hbar^2 N}{10bD'}F\left(\tfrac{1}{2},-1;\tfrac{5}{2};\frac{d_0^2}{b^2}\right)$$

$$+\frac{\pi\gamma^4\hbar^2 N}{10D'}\int_{d_0}^b g(r(0))r(0)^{-2}F\left(\tfrac{3}{2},-1;\tfrac{5}{2};\frac{d_0^2}{r(0)^2}\right)\,\mathrm{d}r(0),$$

and therefore

$$j(0)=\frac{\pi\gamma^4\hbar^2 N}{10D'}\left\{\frac{1}{b}\left(1-\tfrac{1}{5}\frac{d_0^2}{b^2}\right)+\int_{d_0}^b g(r(0))[r(0)^{-2}-\tfrac{3}{5}d_0^2 r(0)^{-4}]\,\mathrm{d}r(0)\right\}.$$

We then deduce from (4.60) that in the extreme narrowing approximation the relaxation rate,

$$\frac{1}{T_1}=\frac{1}{T_2}=\frac{2\pi\gamma^4\hbar^2 NI(I+1)}{3D'}$$

$$\times\left\{\frac{1}{b}\left(1-\frac{d_0^2}{5b^2}\right)+\int_{d_0}^b g(r(0))[r(0)^{-2}-\tfrac{3}{5}d_0^2 r(0)^{-4}]\,\mathrm{d}r(0)\right\}. \tag{5.15}$$

This agrees with equation (12) of Harmon & Muller (1969), if we put in the present section only

$$I_1=\int_{d_0}^b g(r(0))r(0)^{-2}\,\mathrm{d}r(0)$$

$$I_2=\int_{d_0}^b g(r(0))r(0)^{-4}\,\mathrm{d}r(0). \tag{5.16}$$

In the hard sphere approximation (5.15) yields

$$\frac{1}{T_1} = \frac{1}{T_2} = \frac{8\pi N\gamma^4 h^2 I(I+1)}{15 D' d_0}. \tag{5.17}$$

We next examine the situation when $D'\tau\rho^2 \ll 1$. Then (5.14) may be approximated by

$$A(\rho) = \frac{1}{1 + D'\tau\rho^2}, \tag{5.18}$$

which has the mathematical structure of (5.10). From (5.18) and (2.26) it follows that

$$P_1(r) = \frac{\exp\left[-\dfrac{r}{(D'\tau)^{1/2}}\right]}{4\pi D'\tau r}. \tag{5.19}$$

Torrey showed that this probability density function provides a model of a nuclear spin which can exist either bound in a potential well or in a thermally excited state, where the nucleus may be in a rapid diffusive motion. The mean lifetime in the excited state is small compared with the mean lifetime in the trapped state. It is assumed that the diffusive motion does not contribute to the relaxation process.

Let us now calculate $j(0)$ from (5.13) and (5.18). We may express $j(0)$ by

$$j(0) = \frac{3\pi\gamma^4 h^4 N\tau}{5 d_0^{3/2} b^{3/2}} \mathcal{L}_1 + \frac{3\pi\gamma^4 h^2 N\tau}{5 d_0^{3/2}} \mathcal{L}_2, \tag{5.20}$$

where

$$\mathcal{L}_1 = (D'\tau)^{-1} \int_0^\infty J_{3/2}(\rho d_0) J_{3/2}(\rho b)\rho^{-3}\, \mathrm{d}\rho + \int_0^\infty J_{3/2}(\rho d_0) J_{3/2}(\rho b)\rho^{-1}\, \mathrm{d}\rho$$

$$\mathcal{L}_2 = \int_{d_0}^b g(r(0)) r(0)^{-3/2}\, \mathrm{d} r(0) \int_0^\infty [1 + (D'\tau\rho^2)^{-1}] J_{3/2}(\rho d_0) J_{5/2}(\rho r(0))\, \mathrm{d}\rho.$$

Performing the integrations as before we obtain

$$\mathcal{L}_1 = \frac{b^{1/2} d_0^{3/2}}{6 D'\tau}\left(1 - \frac{1}{5}\frac{d_0^2}{b^2}\right) + \frac{d_0^{3/2}}{3 b^{3/2}} \tag{5.21}$$

$$\mathcal{L}_2 = \int_{d_0}^b \left\{ r(0)^{-2}\frac{d_0^{3/2}}{6 D'\tau} + r(0)^{-4} d_0^{3/2} - r(0)^{-4}\frac{d_0^{7/2}}{10 D'\tau} \right\} g(r(0))\, \mathrm{d} r(0).$$

On introducing I_1 and I_2 from (5.16) the last equation becomes

$$\mathcal{L}_2 = \frac{d_0^{3/2}}{6 D'\tau} I_1 + d_0^{3/2} I_2 - \frac{d_0^{7/2}}{10 D'\tau} I_2. \tag{5.22}$$

Then from (2.43) and (5.20)–(5.22)

$$j(0) = \frac{\pi\gamma^4\hbar^2 N\tau}{5d_0^3}\left\{\frac{d_0^3}{b^3} + \frac{3d_0^3}{b\langle r^2\rangle}\left(1 - \frac{d_0^2}{5b^2}\right) + 3d_0^3 I_2 + \frac{3d_0^3}{\langle r^2\rangle}\left(I_1 - \frac{3d_0^2}{5}I_2\right)\right\},$$

and from (4.60) we deduce that in the extreme narrowing approximation

$$\frac{1}{T_1} = \frac{1}{T_2} = \frac{4\pi\gamma^4\hbar^2 NI(I+1)\tau}{3d_0^3}$$

$$\times\left\{\frac{d_0^3}{b^3} + \frac{3d_0^3}{b\langle r^2\rangle}\left(1 - \frac{d_0^2}{5b^2}\right) + 3d_0^3 I_2 + \frac{3d_0^3}{\langle r^2\rangle}\left(I_1 - \frac{3d_0^2}{5}I_2\right)\right\}. \qquad (5.23)$$

This agrees with equation (18) of Harmon & Muller (1969).

For the hard sphere model $b = d_0$, so I_1 and I_2 vanish and (5.23) reduces to

$$\frac{1}{T_1} = \frac{1}{T_2} = \frac{4\pi\gamma^4\hbar^2 NI(I+1)\tau}{3d_0^3}\left\{1 + \frac{12}{5}\frac{d_0^2}{\langle r^2\rangle}\right\}.$$

Multiplying the numerator and denominator of the multiplying factor by D' and using (2.43) we deduce that

$$\frac{1}{T_1} = \frac{1}{T_2} = \frac{8\pi\gamma^4\hbar^2 NI(I+1)}{15D'd_0}\left\{1 + \frac{5}{12}\frac{\langle r^2\rangle}{d_0^2}\right\}. \qquad (5.24)$$

Comparing this result with (5.17), which is a consequence of (5.14), we find that in addition to the previous expression for the relaxation rates due to diffusion there is

$$\frac{2\pi\gamma^4\hbar^2 NI(I+1)\langle r^2\rangle}{9D'd_0}, \qquad (5.25)$$

called the *jump term*. This difference is due to the assumption $D'\tau\rho^2 \ll 1$, which allowed us to replace (5.14) by (5.18). Thence there arose the possibility of the nuclear spin being bound in a potential well related to the probability density function of (5.19).

5.3 Brownian motion models

We examine here how Brownian motion considerations may be applied to the hard sphere model of identical molecules, each of which contains a dipole at its centre. The Langevin equation of translational motion (2.48) with a single frictional coefficient mB' is applicable, and will lead to the conditional probability density w of (2.50). It would be very inconvenient to employ such complicated expression in the calculations, so we agree to work in the Debye approximation (2.58), which leads to (2.52) for w and to the diffusion equation (2.54) with D' defined by (2.53).

We assume that we have a uniform density of N independent spins per unit volume and proceed to calculate $j(\omega)$ from (4.21), replacing p of (5.1) by N and taking w from (2.45):

$$w(\mathbf{r}_j(0)-\mathbf{r}_i(0),\mathbf{R};\mathbf{r}_j(t)-\mathbf{r}_i(t),\mathbf{r})=\frac{\exp\left[-\dfrac{|\mathbf{r}-\mathbf{R}|^2}{8D't}\right]}{(8\pi D't)^{3/2}}.$$

This equation and the result (Peirce, 1929, p. 64)

$$\int_0^\infty e^{-a^2x^2}\cos bx\,dx=\frac{\pi^{1/2}\exp\{-b^2/(4a^2)\}}{2a}\qquad(a>0)$$

allow us to write

$$w(\mathbf{r}_j(0)-\mathbf{r}_i(0),\mathbf{R};\mathbf{r}_j(t)-\mathbf{r}_i(t),\mathbf{r})=\frac{1}{8\pi^3}\int\exp(-2D't\rho^2)\,e^{-i(\boldsymbol{\rho}\cdot\mathbf{R}-\mathbf{r})}\,d^3\rho.$$

Comparing this with (5.2) we see without further investigation that we may obtain $j(\omega)$ from (5.7) by making the replacement $1-A(\rho)\mapsto D'\tau\rho^2$. Thus we now have

$$j(\omega)=\frac{12\pi N\gamma^4\hbar^2 D'}{5d_0^3}\int_0^\infty\frac{\rho[J_{3/2}(\rho d_0)]^2}{\omega^2+4D'^2\rho^4}\,d\rho.\qquad(5.26)$$

Replacing the argument of the Bessel function by x we obtain from (5.26)

$$j(\omega)=\frac{3\pi N\gamma^4\hbar^2}{5D'd_0}\int_0^\infty\frac{[J_{3/2}(x)]^2x\,dx}{x^4+\tfrac14 u^4},\qquad(5.27)$$

where

$$u^2=\frac{d_0^2\omega}{D'}.\qquad(5.28)$$

The integral in (5.27) has been evaluated by Hubbard (1966) and on substitution leads to

$$j(\omega)=\frac{3\pi N\gamma^4\hbar^2}{5D'd_0}\frac{u^2-2+e^{-u}[(u^2-2)\sin u+(u^2+4u+2)\cos u]}{u^5}.(5.29)$$

It is immediately obvious that for $u\gg1$ the function of u in (5.29) tends to u^{-3}. On expanding the function as a series in u it may be shown that for $u\ll1$ the function tends to $2/15$.

The relaxation times may now be deduced from (4.58) and (4.59). From (5.28) and (5.29) we then find that

$$\frac{1}{T_1}=\frac{1}{T_2}=\frac{8\pi N\gamma^4\hbar^2 I(I+1)}{15D'd_0}\qquad\left(\omega_0\ll\frac{D'}{d_0^2}\right).\qquad(5.30)$$

This result is the same as (5.17), though the definitions of D' for the two models are different. The jump term (5.25) vanishes in this model because the mean squared flight distance becomes vanishingly small. If we put $I = \frac{1}{2}$ in (5.30), we obtain

$$\frac{1}{T_1} = \frac{1}{T_2} = \frac{2\pi N\gamma^4\hbar^2}{5D'd_0} \qquad \left(\omega_0 \ll \frac{D'}{d_0^2}\right). \tag{5.31}$$

This agrees with the result of Abragam (1961, p. 302, eq. (115)), where d_0 is replaced by $2a$, that is twice the radius of the molecule.

When $\omega_0 \gg D'/d_0^2$, so that $u \gg 1$,

$$j(\omega_0) = \frac{3\pi N\gamma^4\hbar^2 D'^{1/2}}{5d_0^4\omega_0^{3/2}}$$

$$j(2\omega_0) = \frac{3\pi N\gamma^4\hbar^2 D'^{1/2}}{5d_0^4\omega_0^{3/2}} \times \frac{2^{1/2}}{4}. \tag{5.32}$$

On the other hand from (5.29)

$$j(0) = \frac{2\pi N\gamma^4\hbar^2}{25D'd_0}.$$

We deduce that

$$\frac{1}{T_1} = \frac{4(1+2^{1/2})\pi N\gamma^4\hbar^2 D'^{1/2}I(I+1)}{5d_0^4\omega_0^{3/2}} \qquad \left(\omega_0 \gg \frac{D'}{d_0^2}\right) \tag{5.33}$$

$$\frac{1}{T_2} = \pi\gamma^4\hbar^2 NI(I+1)\left\{\frac{4}{25D'd_0} + \frac{\left(2 + \frac{2^{1/2}}{5}\right)D'}{d_0^4\omega_0^{3/2}}\right\} \qquad \left(\omega_0 \gg \frac{D'}{d_0^2}\right). \tag{5.34}$$

5.4 Other models

The molecular model of the hard sphere with the dipole at its centre is obviously rather crude, and we shall now describe briefly some refinements that have been made in the theory. Hubbard (1963a) investigated the case of spherical molecules of radius a, which contain identical nuclei of spin $\frac{1}{2}$ distributed uniformly at a distance b from the centre. As a result of the eccentric positions of the nuclei the distance between two nuclei in different molecules will depend on both the translational and rotational motions of the two molecules. Consequently the spectral density $j(\omega)$ will have both translational and rotational contributions. Working in the extreme narrowing approximation Hubbard found that the intermolecular interactions gave

$$\frac{1}{T_1} = \frac{1}{T_2} = \frac{\pi\gamma^4\hbar^2 N}{5aD'}\left[1 + 0.233\left(\frac{b}{a}\right)^2 + 0.15\left(\frac{b}{a}\right)^4 + \cdots\right]. \tag{5.35}$$

Since $d_0 = 2a$, we see that the first term of the series produces agreement with (5.31). The remaining terms of the series, which arise from the rotational motion, provide a correction that is probably less than 10 per cent.

Zeidler (1975) generalized the investigations of Hubbard by discarding the assumption of uniform spin distribution and taking account of the non-uniform distribution of interacting nuclei around any given one. This is accomplished by describing the state of the system at time zero by a molecular pair distribution function. In order to simplify lengthy mathematical expressions he restricted the explicit results to a liquid which includes different isotopes for which the structure and dynamics of the liquid remain unchanged but in which the interacting dipolar nuclei are at different distances from the centres of their molecules. A general integral expression for the relaxation time in the extreme narrowing approximation is given in terms of g-coefficients which arise in an expansion of the pair distribution function.

Ayant *et al.* (1975) examined the hard sphere model of two molecules of unequal radii, the interacting nuclei being at the centres. They pointed out that the expression for w given by (2.45) would not hold when the value of r is close to d_0, and they sought to solve the diffusion equation with the boundary condition $\partial w/\partial r = 0$ for $r = d_0$. As a result (5.29) is now altered to

$$j(\omega) = \frac{3\pi N \gamma^4 \hbar^2}{5D'd_0} \frac{\frac{3}{2}u^2 + \frac{15}{2}u + 12}{\frac{1}{8}u^6 + u^5 + 4u^4 + \frac{27}{2}u^2 + 81u + 81}. \tag{5.36}$$

Consequently the factor $2/15$ which appeared for $u \ll 1$ is replaced by $4/27$, so that (5.31) is changed to

$$\frac{1}{T_1} = \frac{1}{T_2} = \frac{4\pi N \gamma^4 \hbar^2}{9D'd_0} \qquad \left(\omega_0 \ll \frac{D'}{d_0^2}\right). \tag{5.37}$$

When $u \gg 1$, (5.36) yields

$$j(\omega) = \frac{36\pi N \gamma^4 \hbar^2 D'}{5d_0^5 \omega^2} \qquad \left(\omega \gg \frac{D'}{d_0^2}\right), \tag{5.38}$$

so that the $\omega^{-3/2}$-dependence of $j(\omega)$, as seen from (5.32), is replaced by a ω^{-2}-dependence.

These investigations were extended by Ayant *et al.* (1977) to the case of spherical molecules having eccentric nuclear spins. Their analysis assumes that the translational and rotational interactions are independent of each other. For the intermolecular interaction of centred and eccentric dipoles they give analytical formulae for spectral densities that lead to values of relaxation times. When the spins are all eccentric and they belong to identical molecules, they provide the same information in the extreme narrowing approximation and for $\omega \gg D'/d_0^2$. In the latter case the ω^{-2}-dependence of (5.38) is still present.

6

Relaxation by intramolecular dipolar interactions

6.1 Intramolecular interactions of like spins

We expressed the longitudinal and transverse relaxation times for systems of like spins in terms of spectral densities by (4.58) and (4.59). The spectral density $j(\omega)$ for intramolecular interactions may be obtained by putting $\gamma_I = \gamma_S = \gamma$ in (4.23), so that

$$j(\omega) = \frac{3\pi\gamma^4\hbar^2}{25r^6} \sum_{n,n'=-2}^{2} Y_{2n'}^*(\theta',\phi')Y_{2n}(\theta',\phi') \int_{-\infty}^{\infty} \langle R^+(t)\rangle_{nn'}\, e^{-i\omega t}\, dt.$$

(6.1)

According to (D.5) and (D.11) of Appendix D we may express (6.1) as

$$j(\omega) = \frac{3\pi\gamma^4\hbar^2}{25r^6} \sum_{n,n'=-2}^{2} Y_{2n}(\theta',\phi')\sigma(\omega)_{nn'}Y_{2n'}^*(\theta',\phi'),$$

(6.2)

where the operator $\sigma(\omega)$ is defined by (D.6) and (D.10). We omit the superscript 2 to $\sigma(\omega)_{nn'}$ in (6.2), since in this chapter we shall employ only the representation of rank 2.

To complete the calculation of relaxation times we need the value of the operator $\langle R^+(t)\rangle$. To find this we choose a theory that is applicable to non-spherical molecules and that produces analytical results. We therefore adopt the rotational Brownian motion theory introduced in subsection 2.4.2. We also accept the assumptions made in subsection 2.4.4. This justifies applying the value of $R(t)$ deduced from the body frame coordinate system with origin at the centre of mass and employed for the Euler–Langevin equations (2.57) to a parallel coordinate system with origin at one of the dipoles involved in the dipole–dipole interaction.

The value of $\langle R(t)\rangle$ and consequently of $\sigma(\omega)_{nn'}$ will depend on the shape of the molecule that contains the two interacting spins. In the following sections we shall study the cases of molecules that are spherical, linear, symmetric and asymmetric tops. The values of $\langle R(t)\rangle$ and $\sigma(\omega)_{nn'}$ have been calculated elsewhere and are collected together in Appendix D, from which we shall now quote.

6.2 Spherical molecules

The value of $\langle R(t)\rangle$ given by (D.20) is a multiple of the identity operator. Hence $\sigma(\omega)_{mn}$ and $\tau(\omega)_{mn}$ defined by (D.6) and (D.10) are multiples of the five-dimensional unit matrix. We therefore deduce from (6.2) that

$$j(\omega) = \frac{3\pi\gamma^4\hbar^2\sigma(\omega)_{00}}{25r^6} \sum_{m=-2}^{2} |Y_{2m}(\theta', \phi')|^2,$$

which simplifies to

$$j(\omega) = \frac{3\gamma^4\hbar^2\sigma(\omega)_{00}}{20r^6} \tag{6.3}$$

(Rose, 1957, eq. (4.28)). Substituting from (D.24) into (6.3), (4.58) and (4.59) we deduce that in the inertial theory

$$\frac{1}{T_1} = \frac{2\gamma^4\hbar^2 I(I+1)}{5r^6 B}\left\{ \frac{G'}{G'^2+\omega_0'^2} + \frac{6\kappa[G'(1+G')-\omega_0'^2]}{(G'^2+\omega_0'^2)[(1+G')^2+\omega_0'^2]} \right.$$

$$\left. + \frac{4G'}{G'^2+4\omega_0'^2} + \frac{24\kappa[G'(1+G')-4\omega_0'^2]}{(G'^2+\omega_0'^2)[(1+G')^2+4\omega_0'^2]} + \cdots \right\}$$

$$\frac{1}{T_2} = \frac{\gamma^4\hbar^2 I(I+1)}{5r^6 B}\left\{ \frac{3}{G'} + \frac{5G'}{G'^2+\omega_2'^0} + \frac{2G'}{G'^2+4\omega_2'^2} + \cdots \right. \tag{6.4}$$

$$+ \frac{18\kappa}{G'(1+G')} + \frac{30\kappa[G'(1+G')-\omega_0'^2]}{(G'^2+\omega_0'^2)[(1+G')^2+\omega_0'^2]}$$

$$\left. + \frac{12\kappa[G'(1+G')-4\omega_0'^2]}{(G'^2+4\omega_0'^2)[(1+G')^2+4\omega_0'^2]} + \cdots \right\}.$$

The notation is explained in Appendix D.

The orientational correlation time τ_2 related to the spherical harmonic $Y_{2m}(\theta(t), \phi(t))$ is, according to (D.25), given by

$$\tau_2 = \tau(0)_{00}.$$

Substituting zero for ω' in (D.23) and expanding G' from (D.22) in powers of the small parameter κ we obtain

$$\tau_2 = \frac{I_1 B}{6kT}(1+\tfrac{11}{2}\kappa-\tfrac{83}{6}\kappa^2+\cdots). \tag{6.5}$$

In analogy with (2.53) for translational motion we define the *rotational diffusion coefficient D* by

$$D = \frac{kT}{I_1 B}. \tag{6.6}$$

We see from (4.19) that $H_q(t)$ is for intramolecular interaction a constant multiple of $Y_{2q}(\theta(t), \phi(t))$. On transforming from the laboratory to the body fixed coordinate system it follows that $H_{2q}(\mathbf{a})$ obeys (D.16) with $l=2$. Hence τ'_2, the correlation time associated with $H_q(t)$, is identical with τ_2.

It may at this stage be worthwhile to recall that since, as was defined in section 4.1, $\theta(t)$ and $\phi(t)$ are the polar and azimuthal angles with respect to the laboratory system of the line joining one dipole to the other, it follows that θ', ϕ' are the polar, azimuthal angles of the same line with reference to the body fixed coordinate axes taken with origin at the first dipole and with directions parallel to the principal axes of inertia through the centre of mass.

From (D.23)–(D.25) we see that

$$\sigma(0)_{mn}=2\tau(0)_{mn}=2\delta_{mn}\tau(0)_{00}=2\delta_{mn}\tau_2,$$

which with (6.3) yields

$$j(0)=\frac{3\gamma^4\hbar^2\tau_2}{10r^6}. \tag{6.7}$$

Hence (4.60) gives for the extreme narrowing approximation

$$\frac{1}{T_1}=\frac{1}{T_2}=\frac{2\gamma^4\hbar^2I(I+1)\tau_2}{r^6}. \tag{6.8}$$

Since $\tau_2=\tau'_2$, we deduce from (D.18) that (6.7) and therefore (6.8) are true also for non-spherical molecules. However, the value of τ_2 will differ for the different molecular models. In the case of the sphere (6.5), (6.6) and (6.8) yield

$$\frac{1}{T_1}=\frac{1}{T_2}=\frac{\gamma^2\hbar^2I(I+1)}{3r^6D}(1+\tfrac{11}{2}\kappa-\tfrac{83}{6}\kappa^2+\cdots). \tag{6.9}$$

In the Debye approximation (6.3) and (D.30) give

$$j(\omega)=\frac{\gamma^4\hbar^2I_1B}{20r^6kT}\frac{1}{1+\left(\dfrac{I_1B\omega}{6kT}\right)^2} \tag{6.10}$$

and we deduce from (4.58), (4.59) and (6.6) that

$$\frac{1}{T_1}=\frac{\gamma^4\hbar^2I(I+1)}{15r^6D}\left\{\frac{1}{1+\left(\dfrac{\omega_0}{6D}\right)^2}+\frac{4}{1+\left(\dfrac{\omega_0}{3D}\right)^2}\right\} \tag{6.11}$$

$$\frac{1}{T_2}=\frac{\gamma^4\hbar^2I(I+1)}{30r^6D}\left\{3+\frac{5}{1+\left(\dfrac{\omega_0}{6D}\right)^2}+\frac{2}{1+\left(\dfrac{\omega_0}{3D}\right)^2}\right\}. \tag{6.12}$$

From (6.5) and (6.6) the correlation time is now $\tfrac{1}{6}D$, and so (6.8) yields

$$\frac{1}{T_1} = \frac{1}{T_2} = \frac{\gamma^4 \hbar^2 I(I+1)}{3r^6 D},$$ (6.13)

as may be verified immediately from (6.9) or from (6.11) and (6.12).

Abragam (1961, pp. 298–300) investigated the problem of a rigid spherical molecule of radius a in Debye approximation by supposing that the probability density function $p(\theta(t), \phi(t), t)$ for finding the dipole–dipole axis in the direction $\theta(t), \phi(t)$ at time t obeys the diffusion equation

$$\frac{\partial p(\theta(t), \phi(t), t)}{\partial t} = D\nabla^2 p(\theta(t), \phi(t), t),$$ (6.14)

which is analogous to (2.54) for translational motion. In (6.14) the value of D is given by (6.6), if we accept the macroscopic relation (Lamb, 1932, p. 589)

$$I_1 B = 8\pi a^3 \eta,$$ (6.15)

where η is the coefficient of viscosity of the environment of the molecule. Defining τ_2 as in (D.31) with $I_1 B$ given by (6.15) Abragam deduces that the autocorrelation functions of $F^{(0)}(t), F^{(1)}(t), F^{(2)}(t)$ of (4.88) depend on time through the factor $\exp[-|t|/\tau_2]$. Consequently their spectral densities are all proportional to $(1 + \omega^2 \tau_2^2)^{-1}$. He gives explicit results for T_1 only and these are in agreement with (6.11) and (6.13). The equivalence of the discussion of this problem by means of a stochastic differential equation or of a rotational diffusion equation is established in the theory of rotational Brownian motion (McConnell, 1980b, section 6.2).

6.3 Linear molecules

When the molecule is linear, the matrices $\tau(\omega)_{mn}, \sigma(\omega)_{mn}$ are diagonal with $\tau(\omega)_{-m,-m} = \tau(\omega)_{mm}, \sigma(\omega)_{-m,-m} = \sigma(\omega)_{mm}$, but the matrices are not just multiples of the unit matrix. Thus we deduce from (6.2) that

$$j(\omega) = \frac{3\pi\gamma^4\hbar^2}{25r^6} \{|Y_{20}(\theta', \phi')|^2\sigma(\omega)_{00} + 2|Y_{21}(\theta', \phi')|^2\sigma(\omega)_{11}$$

$$+ 2|Y_{22}(\theta', \phi')|^2\sigma(\omega)_{22}\}.$$ (6.16)

However, since the line joining the dipoles is also the third coordinate axis of the body frame used in section D.2, θ' vanishes. Then, from (4.8),

$$Y_{20}(\theta', \phi') = \left(\frac{5}{4\pi}\right)^{1/2}, \qquad Y_{21}(\theta', \phi') = Y_{22}(\theta', \phi') = 0$$

and (6.16) reduces to

$$j(\omega) = \frac{3\gamma^4\hbar^2}{20r^6}\,\sigma(\omega)_{00}. \tag{6.17}$$

On substitution from (D.40) (6.17) becomes

$$j(\omega) = \frac{3\gamma^4\hbar^2}{10r^6 B}\left\{\frac{(1+6\kappa+33\kappa^2+\cdots)G_0}{G_0^2+\omega'^2} - \frac{(6\kappa+48\kappa^2+\cdots)(1+G_0)}{(1+G_0)^2+\omega'^2}\right.$$

$$\left. - \frac{12\kappa^2[(1+G_0)^2-\omega'^2]}{[(1+G_0)^2+\omega'^2]^2} + \frac{15\kappa^2(2+G_0)}{(2+G_0)^2+\omega'^2} + \cdots\right\}, \tag{6.18}$$

where from (D.34)

$$G_0 = 6\kappa(1+\kappa+\tfrac{8}{3}\kappa^2+\cdots).$$

The relaxation times T_1 and T_2 may be deduced from (4.58), (4.59) and (6.18).

It is easily found from (6.18) that

$$j(0) = \frac{\gamma^4\hbar^2}{20r^6 D}(1+5\kappa-\tfrac{32}{3}\kappa^2+\cdots) \tag{6.19}$$

with D given by (6.6). Then we deduce from (4.60) and (6.19) the result for the extreme narrowing approximation

$$\frac{1}{T_1} = \frac{1}{T_2} = \frac{\gamma^4\hbar^2 I(I+1)}{3r^6 D}(1+5\kappa-\tfrac{32}{3}\kappa^2+\cdots). \tag{6.20}$$

Since the ratio of T_1^{-1} to τ_2 is independent of the molecular model, (6.8) is true also for the linear model and when combined with (6.20) it yields

$$\tau_2 = \frac{1}{6D}(1+5\kappa-\tfrac{32}{3}\kappa^2+\cdots). \tag{6.21}$$

We may deduce from (6.18)–(6.21) the corresponding results in the rotational diffusion limit by letting κ tend to zero and employing

$$\kappa B = \frac{kT}{I_1 B} = D.$$

We immediately find that the equations for the spherical molecule (6.10)–(6.13) and (D.31) remain true for the linear molecule (McConnell, 1986a). The basic reason for this may be seen by considering the expression in (D.43) for $\langle R(t)\rangle$ and remembering that for the calculation of relaxation and correlation times we need only the matrix element $\langle R^+(t)\rangle_{00}$. From (D.43)

$$\langle R^+(t)\rangle_{00} = \langle R(t)\rangle_{00} = \left\{\exp\left[-\frac{kT}{I_1 B}(J^2-J_3^2)t\right]\right\}_{00}$$

In the five-dimensional representation with the spherical harmonics Y_{2m} as basis the matrix $(J_3^2)_{mn}$ is diagonal with diagonal elements $4, 1, 0, 1, 4$, the zero corresponding to $m = n = 0$ (Edmonds, 1968, p. 17). Hence we have

$$\langle R^+(t)\rangle_{00} = \left\{ \exp\left[-\frac{kT}{I_1 B} J^2 t \right] \right\}_{00},$$

and this is precisely what we would obtain for the spherical molecule, as we see from (D.27).

6.4 Asymmetric molecules

When the molecule is asymmetric, we obtain from (6.2) and (D.53)

$$j(\omega) = \frac{3\pi\gamma^4\hbar^2}{25r^6} [2(A + A^*)|Y_{22}(\theta', \phi')|^2 + 2(B + B^*)|Y_{21}(\theta', \phi')|^2$$

$$+ (C + C^*)|Y_{20}(\theta', \phi')|^2 + (D + D^*)\{Y_{20}(\theta', \phi')[Y_{22}(\theta', \phi')$$

$$+ Y_{2, -2}(\theta', \phi')] + [Y_{22}^*(\theta', \phi') + Y_{2, -2}^*(\theta', \phi')]Y_{20}^*(\theta', \phi')\}$$

$$+ (E + E^*)\{Y_{2, -1}(\theta', \phi')Y_{21}^*(\theta', \phi') + Y_{21}(\theta', \phi')Y_{2, -1}^*(\theta', \phi')\}$$

$$+ (F + F^*)\{Y_{2, -2}(\theta', \phi')Y_{22}^*(\theta', \phi') + Y_{22}(\theta', \phi')Y_{2, -2}^*(\theta', \phi')\}]. \quad (6.22)$$

On substitution from (4.8) this becomes

$$j(\omega) = \frac{3\gamma^4\hbar^2}{80r^6} \{3 \sin^4\theta'(A + A^*) + 3 \sin^2 2\theta'(B + B^*)$$

$$+ (3 \cos^2\theta' - 1)^2(C + C^*) + 2(6^{1/2})(3 \cos^2\theta' - 1) \sin^2\theta' \cos 2\phi'(D + D^*)$$

$$- 3 \sin^2 2\theta' \cos 2\phi'(E + E^*) + 3 \sin^4\theta' \cos 4\phi'(F + F^*)\}. \quad (6.23)$$

The explicit expressions for $j(\omega)$ and consequently for T_1^{-1}, T_2^{-1} deduced from (4.58) and (4.59) will clearly be very lengthy.

In the extreme narrowing approximation we denote by $A(0), B(0), \ldots$ the values of $A, B, \ldots$ when $\omega = 0$, and we see from (D.51) and (D.52) that these values are real. Hence from (6.23)

$$j(0) = \frac{3\gamma^4\hbar^2}{40r^6} \{3 \sin^4\theta' A(0) + 3 \sin^2 2\theta' B(0)$$

$$+ (3 \cos^2\theta' - 1)^2 C(0) + 2(6^{1/2})(3 \cos^2\theta' - 1) \sin^2\theta' \cos 2\phi' D(0)$$

$$- 3 \sin^2 2\theta' \cos 2\phi' E(0) + 3 \sin^4\theta' \cos 4\phi' F(0)\}, \quad (6.24)$$

so that from (4.60) and (6.24)

$$\frac{1}{T_1} = \frac{1}{T_2} = \frac{\gamma^4\hbar^2 I(I+1)}{2r^6} \{3 \sin^4\theta' A(0) + 3 \sin^2 2\theta' B(0)$$

$$+ (3 \cos^2\theta' - 1)^2 C(0) + 2(6^{1/2})(3 \cos^2\theta' - 1) \sin^2\theta' \cos 2\phi' D(0)$$

$$- 3 \sin^2 2\theta' \cos 2\phi' E(0) + 3 \sin^4\theta' \cos 4\phi' F(0)\}. \quad (6.25)$$

Then adopting (6.7) for the asymmetric molecule we obtain from (6.24)

$$\tau_2 = \tfrac{1}{4}\{3 \sin^4 \theta' A(0) + 3 \sin^2 2\theta' B(0) + (3 \cos^2 \theta' - 1)^2 C(0)$$
$$+ 2(6^{1/2})(3 \cos^2 \theta' - 1) \sin^2 \theta' \cos 2\phi' D(0) - 3 \sin^2 2\theta' \cos 2\phi' E(0)$$
$$+ 3 \sin^4 \theta' \cos 4\phi' F(0)\}. \tag{6.26}$$

This shows that there is a continuously infinite number of correlation times.

The results (6.22)–(6.26) can be taken over for the rotational diffusion limit by making the replacements

$$A \mapsto A', \qquad B \mapsto B', \qquad C \mapsto C', \qquad D \mapsto D', \qquad E \mapsto E', \qquad F \mapsto F' \tag{6.27}$$

and employing (D.52) and (D.57).

As an example of the asymmetric rotator model let us take a planar molecule, the third molecular axis being perpendicular to the plane. Then $\theta' = \pi/2$ and (6.23) reduces to

$$j(\omega) = \frac{3\gamma^4 \hbar^2}{80 r^6} \{3(A + A^*) + (C + C^*)$$
$$- 2(6^{1/2}) \cos 2\phi' (D + D^*) + 3 \cos 4\phi' (F + F^*)\}. \tag{6.28}$$

For simplicity we work in the rotational diffusion limit and extreme narrowing approximation, so that

$$A + A^* = 2A', \qquad C + C^* = 2C'$$
$$D + D^* = 2D', \qquad F + F^* = 2F'. \tag{6.29}$$

From (D.52) and (D.57)

$$A' = \frac{ac - d^2}{a(ac - 2d^2)}, \qquad C' = \frac{a}{ac - 2d^2}$$

$$D' = -\frac{dC'}{a}, \qquad F = \left(\frac{d}{a}\right)^2 C' \tag{6.30}$$

$$a = D_1 + D_2 + 4D_3, \qquad c = 3(D_1 + D_2), \qquad d = (\tfrac{3}{2})^{1/2}(D_1 - D_2).$$

On substituting (6.29) and (6.30) into (6.28) one easily finds that

$$j(0) = \frac{3\gamma^4 \hbar^2}{40 r^6} \left\{ 3A' + C' \left[1 + 2(6)^{1/2} \frac{d}{a} \cos 2\phi' + 3\left(\frac{d}{a}\right)^2 \cos 4\phi' \right] \right\} \tag{6.31}$$

$$A' = \frac{1}{3(D_3 + D_S)} + \frac{(D_1 - D_2)^2}{8D_R(D_3 + D_S)}, \qquad C' = \frac{3(D_3 + D_S)}{4D_R}, \tag{6.32}$$

where we introduce the notation

$$D_S = \tfrac{1}{3}(D_1 + D_2 + D_3)$$
$$D_R = 3(D_2 D_3 + D_3 D_1 + D_1 D_2). \tag{6.33}$$

Expressing $\cos 2\phi'$ and $\cos 4\phi'$ by

$$\cos 4\phi' = 1 - 2\sin^2\phi'$$

$$\cos 4\phi' = 1 - 8\sin^2\phi' + 8\sin^4\phi', \tag{6.34}$$

substituting from (6.32) and (6.33) into (6.31), and ordering the terms in powers of $\sin\phi'$ we obtain

$$j(0) = \frac{9\gamma^4\hbar^2}{40r^6 D_R}\left\{(D_1 + D_S) - \left[(D_1 - D_2) + \frac{(D_1 - D_2)^2}{D_3 + D_S}\right]\sin^2\phi'\right.$$

$$\left. + \frac{(D_1 - D_2)^2}{D_3 + D_S}\sin^4\phi'\right\}. \tag{6.35}$$

We then deduce from (4.60) and (6.35) that

$$\frac{1}{T_1} = \frac{1}{T_2} = \frac{3\gamma^4\hbar^2 I(I+1)}{2r^6 D_R}\left\{(D_1 + D_S) - \left[(D_1 - D_2) + \frac{(D_1 - D_2)^2}{D_3 + D_S}\right]\sin^2\phi'\right.$$

$$\left. + \frac{(D_1 - D_2)^2}{D_3 + D_S}\sin^4\phi'\right\}. \tag{6.36}$$

By employing (6.34) it is easily checked that (6.36) agrees with the result of Huntress (1968). Moreover (6.7) and (6.35) yield

$$\tau_2 = \frac{3}{4D_R}\left\{(D_1 + D_S) - \left[(D_1 - D_2) + \frac{(D_1 - D_2)^2}{D_3 + D_S}\right]\sin^2\phi' + \frac{(D_1 - D_2)^2}{D_3 + D_S}\sin^4\phi'\right\}. \tag{6.37}$$

For the special case where the molecule is modelled as a circular plate $D_2 = D_1$, and (6.33) gives

$$D_S = \tfrac{1}{3}(2D_1 + D_3), \qquad D_R = 3D_1(D_1 + 2D_3). \tag{6.38}$$

We deduce from (6.35)–(6.38) that

$$j(0) = \frac{\gamma^4\hbar^2(5D_1 + D_3)}{40r^6 D_1(D_1 + 2D_3)} \tag{6.39}$$

$$\frac{1}{T_1} = \frac{1}{T_2} = \frac{\gamma^4\hbar^2 I(I+1)(5D_1 + D_3)}{6r^6 D_1(D_1 + 2D_3)} \tag{6.40}$$

$$\tau_2 = \frac{5D_1 + D_3}{12D_1(D_1 + 2D_3)}. \tag{6.41}$$

Comparing (6.39)–(6.41) with (6.10), (6.13) and (D.31) we see that the values for the circular plate are

$$\frac{5D_1 + D_3}{2(D_1 + 2D_3)}$$

times the respective values for the sphere. The multiplying factor can vary between 1/4 and 5/2.

6.5 Symmetric molecules

When the molecule is a symmetric rotator, we put $I_2 = I_1$, $B_2 = B_1$ in (2.57) and (D.49) and we also put $D_2 = D_1$, where D_i has been introduced into section D.3 of Appendix D. Then, from (D.59) and (6.23),

$$j(\omega) = \frac{9\gamma^4\hbar^2}{20r^6} \left\{ \frac{\sin^4 \theta' (D_1 + 2D_3)\left(1 + \frac{2D_1}{B_1} + \frac{4D_3}{B_3}\right)}{(2D_1 + 4D_3)^2 + \omega^2} \right.$$

$$+ \frac{2 \sin^2 \theta' \cos^2 \theta' (5D_1 + D_3)\left(1 + \frac{5D_1}{B_1} + \frac{D_3}{B_3}\right)}{(5D_1 + D_3)^2 + \omega^2}$$

$$\left. + \frac{(3 \cos^2 \theta' - 1)^2 D_1\left(1 + \frac{6D_1}{B_1}\right)}{36D_1^2 + \omega^2} \right\}. \qquad (6.42)$$

The relaxation times T_1, T_2 may be deduced by applying (4.58) and (4.59) to (6.42).

For the extreme narrowing case the last equation gives

$$j(0) = \frac{9\gamma^4\hbar^2}{20r^6} \left\{ \frac{\sin^4 \theta' \left(1 + \frac{2D_1}{B_1} + \frac{4D_3}{B_3}\right)}{4(D_1 + 2D_3)} \right.$$

$$+ \frac{2 \sin^2 \theta' \cos^2 \theta' \left(1 + \frac{5D_1}{B_1} + \frac{D_3}{B_3}\right)}{5D_1 + D_3}$$

$$\left. + \frac{(3 \cos^2 \theta' - 1)^2 \left(1 + \frac{6D_1}{B_1}\right)}{36D_1} \right\}. \qquad (6.43)$$

Then, from (4.60),

$$\frac{1}{T_1} = \frac{1}{T_2} = \frac{3\gamma^4\hbar^2 I(I+1)}{r^6} \left\{ \frac{\sin^4 \theta' \left(1 + \frac{2D_1}{B_1} + \frac{4D_3}{B_3}\right)}{4(D_1 + 2D_3)} \right.$$

$$+ \frac{2 \sin^2 \theta' \cos^2 \theta' \left(1 + \frac{5D_1}{B_1} + \frac{D_3}{B_3}\right)}{5D_1 + D_3}$$

$$+\frac{(3\cos^2\theta'-1)^2\left(1+\dfrac{6D_1}{B_1}\right)}{36D_1}\Bigg\}. \tag{6.44}$$

By applying (6.7) to the present model, as we may, we obtain the correlation time

$$\tau_2=\frac{3}{2}\Bigg\{\frac{\sin^4\theta'\left(1+\dfrac{2D_1}{B_1}+\dfrac{4D_3}{B_3}\right)}{4(D_1+2D_3)}+\frac{2\sin^2\theta'\cos^2\theta'\left(1+\dfrac{5D_1}{B_1}+\dfrac{D_3}{B_3}\right)}{5D_1+D_3}$$

$$+\frac{(3\cos^2\theta'-1)^2\left(1+\dfrac{6D_1}{B_1}\right)}{36D_1}\Bigg\}. \tag{6.45}$$

In the rotational diffusion limit

$$D_i=\frac{kT}{I_iB_i}\qquad(i=1,2,3), \tag{6.46}$$

the correction of relative order κ vanishing according to (2.61). Similarly D_i/B_i vanishes. Hence we obtain from (6.42) and (6.45)

$$j(\omega)=\frac{9\gamma^4\hbar^2}{20r^6}\Bigg\{\frac{\sin^4\theta'(D_1+2D_3)}{(2D_1+4D_3)^2+\omega^2}+\frac{2\sin^2\theta'\cos^2\theta'(5D_1+D_3)}{(5D_1+D_3)^2+\omega^2}$$

$$+\frac{(3\cos^2\theta'-1)^2D_1}{36D_1^2+\omega^2}\Bigg\} \tag{6.47}$$

$$\tau_2=\frac{3}{2}\Bigg\{\frac{\sin^4\theta'}{4(D_1+2D_3)}+\frac{2\sin^2\theta'\cos^2\theta'}{5D_1+D_3}+\frac{(3\cos^2\theta'-1)^2}{36D_1}\Bigg\}. \tag{6.48}$$

There is no reason why θ' should vanish in the present model. If it does, so that the line of dipoles is parallel to the axis of rotational symmetry of the molecule, (6.46)–(6.48) give

$$j(\omega)=\frac{\gamma^4\hbar^2}{20r^6D_1}\frac{1}{1+\left(\dfrac{\omega}{6D_1}\right)^2} \tag{6.49}$$

$$\tau_2=\frac{1}{6D_1}=\frac{I_1B_1}{6kT}. \tag{6.50}$$

Equation (6.49) is the same as (6.10) for the spherical molecule, so we shall obtain the relaxation times from (6.11) and (6.12). Similarly we see that (6.5) and (6.21) with $\kappa=0$ provide the value of τ_2 in (6.50). The reason for these results may be found by extending slightly the argument given at the end of section 6.3 for the linear molecule (McConnell, 1986a). Equation

(D.49) applied to the symmetric rotator in Debye approximation reduces to

$$\langle R(t) \rangle = \exp\{ -[D_1(J_1^2 + J_2^2) + D_3 J_3^2]t \}$$
$$= \exp\{ -[D_1 J^2 + (D_3 - D_1)J_3^2]t \},$$

where D_1 and D_3 are given by (6.46). When $\theta' = 0$, we need only $\langle R(t) \rangle_{00}$ to calculate $j(\omega)$ and τ_2. Since $(J_3^2)_{00}$ vanishes, we have the same results as for the spherical molecule. This will no longer be true, if we work in the inertial theory.

In the extreme narrowing approximation and rotational diffusion limit we deduce from (6.44) that

$$\frac{1}{T_1} = \frac{1}{T_2} = \frac{\gamma^4 \hbar^2 I(I+1)}{2r^6}$$

$$\times \left\{ \frac{3 \sin^4 \theta'}{2D_1 + 4D_3} + \frac{12 \sin^2 \theta' \cos^2 \theta'}{5D_1 + D_3} + \frac{(3 \cos^2 \theta' - 1)^2}{6D_1} \right\}. \quad (6.51)$$

This is also expressible as

$$\frac{1}{T_1} = \frac{1}{T_2} = \frac{\gamma^4 \hbar^2 I(I+1)}{3r^6 D_1} \left[1 + \frac{3(D_1 - D_3)}{5D_1 + D_3} \sin^2 \theta' \left(1 + \frac{3(D_1 - D_3)}{2(D_1 + 2D_3)} \sin^2 \theta' \right) \right]$$

$$(6.52)$$

in agreement with the result of Huntress (1968).

6.6 General observations

The common orientational correlation times and expressions for relaxation rates provide a theoretical justification in rotational diffusion theory for applying Abragam's results to linear and symmetric top models, if in the latter model the line of dipoles is parallel to the axis of symmetry of the molecule. We see from (6.8), which holds for all molecular models and for both inertial and rotational diffusion theory, that there is a simple relation between relaxation and correlation times in the extreme narrowing approximation.

On comparing (6.47) with (6.42) it is seen that the inclusion of inertial effects produces a correction of relative order κ to $j(\omega)$. From experiments on dielectric absorption (Herzfeld, 1964; Leroy, Constant & Desplanques, 1967) it appears that this correction is at most of order 1%. A correction of the same order of magnitude will be produced in the relaxation rates T_1^{-1} and T_2^{-1} deduced from (4.58) and (4.59) and in the correlation time τ_2, as we see, for example, from (6.5) and (6.50).

Let us now consider the condition for the extreme narrowing

approximation. This means that we may replace $j(\omega)$ by $j(0)$ when calculating relaxation times. The study of the symmetric molecules shows that the condition on ω_0 is

$$\omega_0 \ll \frac{kT}{I_1 B_1},\tag{6.53}$$

as is immediately obvious from (6.49). We reported in section D.3 that for inertial calculations D_i differs from $kT/I_i B_i$ by a correction of relative order κ, which is negligible. Now in (6.22) the angular velocity enters through $A, B, \ldots, F, A^*, B^*, \ldots, F^*$, and we see from (D.51) and (D.52) that ω may be neglected, if it is very small compared with D_1, D_2 and D_3. We therefore generalize (6.53) and express the conditions required for the extreme narrowing approximation as

$$\omega_0 \ll \frac{kT}{I_i B_i} \qquad (i = 1, 2, 3).\tag{6.54}$$

In order to estimate whether this condition is reasonable we consider a chloroform molecule $CHCl_3$, which is almost spherical. According to the dielectric theory of Debye (1929) relaxation occurs with a relaxation time τ_D given by

$$\tau_D = \frac{I_1 B}{2kT},$$

so that

$$\frac{kT}{I_1 B} = \frac{1}{2\tau_D}.$$

For pure liquid chloroform at 25°C the experimental value of τ_D is 6.36×10^{-12} s (Goulon *et al.*, 1973) and so $kT/I_1 B$ is 7.86×10^{10} s^{-1}. If the relaxing nucleus in $CHCl_3$ is the proton, then (Andrew, 1969, Appendix 2)

$$\omega_0 = 2\pi \times 42.578 \times 10^4 \text{ s}^{-1} = 2.675 \times 10^6 \text{ s}^{-1},$$

which is four orders of magnitude less than $kT/I_1 B$.

When investigating intermolecular interactions, we made in (5.30) the approximation $\omega_0 \ll D' d_0^{-2}$, where D' is the self-diffusion coefficient and d_0 is twice the radius a of the spherical molecule. It will be of some interest to compare $D' d_0^{-2}$ with $kT/(I_1 B)$. According to (2.53)

$$\frac{D'}{d_0^2} = \frac{kT}{4\zeta_t a^2},$$

where we have written ζ_t for the coefficient of the translational friction mB'. If we accept the Rayleigh–Stokes macroscopic laws (Batchelor, 1967, p. 233; Lamb, 1932, p. 589) relating the coefficients of translational and rotational friction with the viscosity η

$$\zeta_t = 6\pi a \eta, \qquad \zeta_r = 8\pi a^3 \eta, \tag{6.55}$$

we have

$$\frac{D'}{d_0^2} = \frac{kT}{4a^2\zeta_t} = \frac{kT}{24\pi a^3 \eta} = \frac{kT}{3\zeta_r} = \frac{kT}{3I_1 B}.$$

Thus we may say that the condition in (5.30) is for spherical molecules essentially equivalent to the condition (6.54) which defines the extreme narrowing approximation, if it is assumed that

$$\zeta_r = \frac{4a^2}{3}\zeta_t. \tag{6.56}$$

This is a slightly less stringent condition than (6.55). The validity of (6.56) will be discussed in section 11.1.

6.7 Intramolecular interactions of unlike spins

For the case of unlike spins we return to (4.23). Comparing this with (6.1) we see that we may take over all the expressions for $j(\omega)$ given in the previous section for the various molecular models by making the substitution $\gamma^4 \mapsto \gamma_I^2 \gamma_S^2$. However, to obtain relaxation rates we must employ (4.78)–(4.86) and (4.90) in place of (4.56)–(4.60). For unlike spins the rotational diffusion result (6.49) yields

$$j(\omega) = \frac{3\gamma_I^2\gamma_S^2\hbar^2\tau_2}{10r^6(1+\omega^2\tau_2^2)} \tag{6.57}$$

with τ_2 still defined by (6.50). We then deduce that

$$\frac{1}{T_1^{II}} = \frac{2\gamma_I^2\gamma_S^2\hbar^2 S(S+1)\tau_2}{15r^6}$$

$$\times \left\{ \frac{3}{1+\omega_I^2\tau_2^2} + \frac{1}{1+(\omega_I-\omega_S)^2\tau_2^2} + \frac{6}{1+(\omega_I+\omega_S)^2\tau_2^2} \right\}$$

$$\frac{1}{T_1^{IS}} = \frac{2\gamma_I^2\gamma_S^2\hbar^2 I(I+1)\tau_2}{15r^6}$$

$$\times \left\{ -\frac{1}{1+(\omega_I-\omega_S)^2\tau_2^2} + \frac{6}{1+(\omega_I+\omega_S)^2\tau_2^2} \right\} \tag{6.59}$$

$$\frac{1}{T_2^I} = \frac{\gamma_I^2 \gamma_S^2 \hbar^2 S(S+1)\tau_2}{15r^6}$$

$$\times \left\{ 4 + \frac{1}{1+(\omega_I - \omega_S)^2 \tau_2^2} + \frac{3}{1+\omega_I^2 \tau_2^2} \right.$$

$$\left. + \frac{6}{1+\omega_S^2 \tau_2^2} + \frac{6}{1+(\omega_I + \omega_S)^2 \tau_2^2} \right\}. \tag{6.60}$$

The values of T_1^{SS}, T_1^{SI}, T_2^S are obtained by making in (6.58)–(6.60) the substitution

$$I \mapsto S, \qquad S \mapsto I, \qquad \omega_I \mapsto \omega_S, \qquad \omega_S \mapsto \omega_I.$$

In the extreme narrowing approximation (6.58)–(6.60) become

$$\frac{1}{T_1^{II}} = \frac{4\gamma_I^2 \gamma_S^2 \hbar^2 S(S+1)\tau_2}{3r^6} \tag{6.61}$$

$$\frac{1}{T_1^{IS}} = \frac{2\gamma_I^2 \gamma_S^2 \hbar^2 I(I+1)\tau_2}{3r^6} \tag{6.62}$$

$$\frac{1}{T_2^I} = \frac{4\gamma_I^2 \gamma_S^2 \hbar^2 S(S+1)\tau_2}{3r^6}. \tag{6.63}$$

7

Relaxation by scalar interaction

7.1 The scalar interaction

We explain in Appendix C the general method of constructing the components of an irreducible spherical tensor of rank j from the components of two irreducible spherical tensors of ranks j_1, j_2. We now take $j_1 = j_2 = 1, j = 0$, so that the final tensor will have only one component. On putting

$$U_q^1 = I_q, \qquad T_{q'}^1 = S_{q'}$$

into (C.18) we find that the single tensor component is the rotational invariant $-3^{-1/3}(\mathbf{I} \cdot \mathbf{S})$, which is also expressible by (4.13) as

$$-3^{-1/2}(I_0 S_0 + \tfrac{1}{2}[I_+ S_- + I_- S_+]).$$

Let us therefore examine the scalar interaction with the Hamiltonian $\hbar G(t)$ defined by either of the equations

$$\hbar G(t) = \hbar A(\mathbf{I} \cdot \mathbf{S}) \tag{7.1}$$

$$\hbar G(t) = \hbar A(I_0 S_0 + \tfrac{1}{2}[I_+ S_- + I_- S_+]). \tag{7.2}$$

Since the Hamiltonian is self-adjoint, A is real. The time dependence of the Hamiltonian may be in A or in one of the spins.

If I and S are like spins, we have from (4.25)

$$\hbar \mathcal{H}_0 = -\hbar \omega_0 (I_z + S_z). \tag{7.3}$$

It is easily verified that $\hbar \mathcal{H}_0$ commutes with $\hbar G(t)$ as given in (7.1). Then the eigenfunctions of $\hbar \mathcal{H}_0$ are also eigenfunctions of $\hbar G(t)$, and on using them as a basis it is found that the matrix element $\hbar G(t)_{mn}$ vanishes unless $m = n$. Employing this result in the perturbation formalism (McConnell, 1960, p. 158) we see that there are no transitions between different states and therefore no relaxation. When the spins are unlike, (7.3) is replaced as in (4.62) by

$$\hbar \mathcal{H}_0 = -\hbar(\omega_I I_z + \omega_S S_z)$$

and $\hbar \mathcal{H}_0$ no longer commutes with $\hbar G(t)$.

We shall therefore examine relaxation by the scalar interaction of two unlike spins. Let us first consider the Hamiltonian of (7.2) when I and S are time independent and A is time dependent. On comparing (7.2) with (3.38) we see that $H_m(t) \equiv A(t)$ and therefore, from (3.48), that

$$j(\omega) = \frac{1}{2} \int_{-\infty}^{\infty} \langle A^*(0)A(t)\rangle \, e^{-i\omega t} \, dt. \tag{7.4}$$

When applying (3.72) to the calculation of relaxation times T_1, T_2 we may use results deduced in section 4.3 from the element $A^{(0)}$ that appears in the interaction Hamiltonian of (4.17). From our previous discussion we know that the only $j(\omega)$-functions that will result from the spin dependent part of $\hbar G(t)$ are $j(0)$ and $j(\omega_I - \omega_S)$. Since we had in (4.27)

$$A^{(0)} = \frac{4}{6^{1/2}} I_0 S_0 - \frac{1}{6^{1/2}} (I_+ S_- + I_- S_+),$$

we see on comparing with (7.2) that the coefficients of $j(0)$ are now $(6^{1/2}/4)^2$ times those appearing in (4.81)–(4.86), and that the coefficients of $j(\omega_I - \omega_S)$ will be multiplied by a factor $(6^{1/2}/2)^2$. The $j(\omega)$-functions given by (7.4) will, of course, differ from those given by (4.20). Thus for the scalar interaction (4.78)–(4.80) will be true with

$$\frac{1}{T_1^{II}} = \tfrac{2}{3}S(S+1)j(\omega_I - \omega_S), \qquad \frac{1}{T_1^{SS}} = \tfrac{2}{3}I(I+1)j(\omega_I - \omega_S)$$

$$\frac{1}{T_1^{IS}} = -\tfrac{2}{3}I(I+1)j(\omega_I - \omega_S), \qquad \frac{1}{T_1^{SI}} = -\tfrac{2}{3}S(S+1)j(\omega_I - \omega_S)$$

$$\frac{1}{T_2^{I}} = \tfrac{1}{3}S(S+1)\{j(0) + j(\omega_I - \omega_S)\}$$

$$\frac{1}{T_2^{S}} = \tfrac{1}{3}I(I+1)\{j(0) + j(\omega_I - \omega_S)\}.$$

$$\tag{7.5}$$

In the next section we shall employ the above results for a particular value of $A(t)$. The case of a time dependent S will be discussed subsequently.

7.2 Scalar relaxation by chemical exchange

Ramsey & Purcell (1952) drew attention to a mechanism for scalar interaction based on the magnetic interaction between an atomic nucleus and an electron spin of its own atom together with the exchange coupling of the electron spins of two atoms in the same molecule. They considered for simplicity a diatomic molecule in a $^1\Sigma$-state. On account of the Boltzmann probability density function, the magnetic interaction between a nucleus and a spin in the same atom will tend to make the spin of the electron antiparallel to the spin of the nucleus. However, the two electron spins in the molecule are antiparallel, and so an interaction between the two nuclear spins is produced as a second-order effect. This idea was further elaborated for diatomic molecules whose nuclei have spin $\frac{1}{2}$ by Solomon (1955), and by Solomon & Bloembergen (1956), in particular with relation to the hydrogen fluoride molecule which consists of H and ^{19}F.

When the spin I of a nucleus is equal to $\frac{1}{2}$, we can express the spin components in terms of the Pauli spin operators $\sigma_x, \sigma_y, \sigma_z$ by (McConnell, 1960, section 23)

$$I_x = \tfrac{1}{2}\sigma_x, \qquad I_y = \tfrac{1}{2}\sigma_y, \qquad I_z = \tfrac{1}{2}\sigma_z,$$

so that

$$I_+ = \tfrac{1}{2}(\sigma_x + i\sigma_y), \qquad I_- = \tfrac{1}{2}(\sigma_x - i\sigma_y).$$

In two-dimensional representation

$$I_+ = \begin{bmatrix} 0 & 1 \\ 0 & 0 \end{bmatrix}, \qquad I_- = \begin{bmatrix} 0 & 0 \\ 1 & 0 \end{bmatrix}, \qquad I_z = \frac{1}{2}\begin{bmatrix} 1 & 0 \\ 0 & -1 \end{bmatrix}.$$

Hence, if the spin $\frac{1}{2}$ wave functions corresponding to the eigenvalues $\frac{1}{2}, -\frac{1}{2}$ of I_z are $|\frac{1}{2}\rangle, |-\frac{1}{2}\rangle$, respectively, we have

$$I_z|\tfrac{1}{2}\rangle = \tfrac{1}{2}|\tfrac{1}{2}\rangle, \qquad I_+|\tfrac{1}{2}\rangle = 0, \qquad I_-|\tfrac{1}{2}\rangle = |-\tfrac{1}{2}\rangle$$
$$I_z|-\tfrac{1}{2}\rangle = -\tfrac{1}{2}|-\tfrac{1}{2}\rangle, \qquad I_+|-\tfrac{1}{2}\rangle = |\tfrac{1}{2}\rangle, \qquad I_-|-\tfrac{1}{2}\rangle = 0.$$

Employing an obvious notation we have for a two-nucleus system

$$(I_+S_- + I_-S_+)|\tfrac{1}{2}\rangle_I|-\tfrac{1}{2}\rangle_S = |-\tfrac{1}{2}\rangle_I|\tfrac{1}{2}\rangle_S$$
$$(I_+S_- + I_-S_+)|-\tfrac{1}{2}\rangle_I|\tfrac{1}{2}\rangle_S = |\tfrac{1}{2}\rangle_I|-\tfrac{1}{2}\rangle_S.$$

Thus the part $\tfrac{1}{2}hA(t)(I_+S_- + I_-S_+)$ of $hG(t)$ produces a simultaneous spin flip of the two nuclei and consequently chemical exchange.

Let us employ (7.1) to study relaxation by chemical exchange. The function $A(t)$ will have a constant real value A when the two nuclei are in the same molecule and zero otherwise, and so the time dependence of $A(t)$ results from the probability of the two nuclei being in the same molecule. Fixing our attention on a molecule containing a nucleus with spin I we suppose that at time zero a nucleus with spin S is in the same molecule. Let us assume that the probability that the same spin S nucleus is in the molecule at a later time t is expressible as e^{-t/τ_e}. We call τ_e the *chemical exchange time*. Then we have

$$\langle A^*(0)A(t)\rangle = \langle A(0)A(t)\rangle = A^2 e^{-t/\tau_e}.$$

Since $A(t)$ is real and scalar, and since we presume that the system is in a steady state, the correlation function is even. We therefore have for positive or negative t

$$\langle A^*(0)A(t)\rangle = A^2 e^{-|t|/\tau_e}, \tag{7.6}$$

and this shows that the correlation time for $A(t)$ is τ_e. We deduce from (7.4) and (7.6) that

$$j(\omega) = \tfrac{1}{2}A^2 \int_{-\infty}^{\infty} e^{-|t|/\tau_e}(\cos \omega t - i \sin \omega t)\, dt$$

$$= A^2 \,\mathrm{Re}\int_{0}^{\infty} \exp[-t(i\omega + \tau_e^{-1})]\, dt,$$

that is

$$j(\omega) = \frac{A^2\tau_e}{1+\omega^2\tau_e^2}. \tag{7.7}$$

On substitution of (7.7) into (7.5) we deduce that

$$\frac{1}{T_1^{II}} = -\frac{1}{T_1^{SI}} = \frac{\tfrac{2}{3}A^2 S(S+1)\tau_e}{1+(\omega_I-\omega_S)^2\tau_e^2}$$

$$\frac{1}{T_1^{SS}} = -\frac{1}{T_1^{IS}} = \frac{\tfrac{2}{3}A^2 I(I+1)\tau_e}{1+(\omega_I-\omega_S)^2\tau_e^2}$$

$$\frac{1}{T_2^{I}} = \tfrac{1}{3}A^2 S(S+1)\left\{\tau_e + \frac{\tau_e}{1+(\omega_I-\omega_S)^2\tau_e^2}\right\} \tag{7.8}$$

$$\frac{1}{T_2^{S}} = \tfrac{1}{3}A^2 I(I+1)\left\{\tau_e + \frac{\tau_e}{1+(\omega_I-\omega_S)^2\tau_e^2}\right\}.$$

Since the calculation of $j(\omega)$ was based entirely on (7.6), the validity of (7.8) is not restricted to nuclei with spin $\tfrac{1}{2}$.

The spin flip of two nuclei comes not only from (7.2) but also from the $q=0$ part of the dipolar interaction Hamiltonian in (4.17). On taking the value of $A^{(0)}$ from (4.27) we see that the sum of the spin flip contributions to $\hbar G(t)$ is

$$\hbar\{\tfrac{1}{4}\gamma_I\gamma_S\hbar r^{-3}(3\cos^2\theta(t)-1)+\tfrac{1}{2}A(t)\}(I_+S_- + I_-S_+), \tag{7.9}$$

where we have put $r(t)$ equal to r because the interaction is intramolecular, and we have substituted for $Y_{20}(\theta(t), \phi(t))$ from (4.8). In (7.9) the quantity $A(t)$ is unchanged by molecular rotation, so it cannot influence the contribution to relaxation by the dipolar interaction, which arises from the rotational motion. Accordingly when calculating relaxation effects resulting from nuclear spin flips we may add together the contributions from the chemical exchange and dipolar parts of the interaction.

As a preparation for comparing the relative importance of these contributions let us first write (4.78), (4.79) and (4.81)–(4.84) as

$$\frac{d\langle I_z\rangle}{dt} = -\rho\{\langle I_z\rangle - I_0\} - \sigma\{\langle S_z\rangle - S_0\}$$

$$\frac{d\langle S_z\rangle}{dt} = -\rho'\{\langle S_z\rangle - S_0\} - \sigma'\{\langle I_z\rangle - I_0\},$$

(7.10)

where

$$\rho = S(S+1)\{\tfrac{4}{9}j(\omega_I - \omega_S) + \tfrac{4}{3}j(\omega_I) + \tfrac{8}{3}j(\omega_I + \omega_S)\}$$

$$\sigma = I(I+1)\{-\tfrac{4}{9}j(\omega_I - \omega_S) + \tfrac{8}{3}j(\omega_I + \omega_S)\}$$

$$\rho' = I(I+1)\{\tfrac{4}{9}j(\omega_I - \omega_S) + \tfrac{4}{3}j(\omega_S) + \tfrac{8}{3}j(\omega_I + \omega_S)\}$$

$$\sigma' = S(S+1)\{-\tfrac{4}{9}j(\omega_I - \omega_S) + \tfrac{8}{3}j(\omega_I + \omega_S)\}.$$

(7.11)

The values of $j(\omega)$ for different molecular models and in rotational Brownian motion theory were discussed at length in the previous chapter. In the simplest case of a spherical model and with inertial effects neglected we deduce on replacing γ^4 by $\gamma_I^2\gamma_S^2$ in (6.10) that

$$j(\omega) = \frac{3\gamma_I^2\gamma_S^2\hbar^2}{10r^6}\frac{\tau_2}{1 + \omega^2\tau_2^2},$$

(7.12)

where

$$\tau_2 = \frac{I_1 B}{6kT}.$$

In the extreme narrowing approximation (7.12) becomes

$$j(0) = \tfrac{3}{10}\gamma_I^2\gamma_S^2\hbar^2 r^{-6}\tau_2$$

(7.13)

and (7.11) become

$$\rho = \tfrac{4}{3}S(S+1)\delta, \qquad \sigma = \tfrac{2}{3}I(I+1)\delta$$

$$\rho' = \tfrac{4}{3}I(I+1)\delta, \qquad \sigma' = \tfrac{2}{3}S(S+1)\delta,$$

(7.14)

where we have put

$$\delta = \gamma_I^2\gamma_S^2\hbar^2 r^{-6}\tau_2.$$

(7.15)

Similarly we have from (4.85), (4.86), (7.13) and (7.15) that

$$\frac{1}{T_2^I} = \tfrac{4}{3}S(S+1)\delta, \qquad \frac{1}{T_2^S} = \tfrac{4}{3}I(I+1)\delta$$

(7.16)

for a spherical molecule and in the extreme narrowing approximation.

We unite these results with (7.8) to obtain relaxation times for combined dipolar and scalar interactions. The time τ_2 for liquids is usually of order 10^{-11} s (Hasted, 1973, p. 47; McConnell, 1980b, p. 245), whereas τ_e is of order 10^{-6} s (Solomon & Bloembergen, 1956). Taking a magnetic field H_0 having a flux density 10^{-1} T and γ equal to the gyromagnetic ratio 2.675×10^8 s^{-1} T^{-1} of the proton in H_2O we have $\gamma H_0 = 2.675 \times 10^7$ s^{-1}.

Hence we should retain $\omega_I \tau_e$, $\omega_S \tau_e$ while neglecting $\omega_I \tau_2$, $\omega_S \tau_2$ in the extreme narrowing approximation. Then we have (7.10) with, from (7.8) and (7.14),

$$\rho = S(S+1)\left\{ \tfrac{4}{3}\delta + \tfrac{2}{3}A^2 \frac{\tau_e}{1+(\omega_I-\omega_S)^2\tau_e^2} \right\}$$

$$\sigma = I(I+1)\left\{ \tfrac{2}{3}\delta - \tfrac{2}{3}A^2 \frac{\tau_e}{1+(\omega_I-\omega_S)^2\tau_e^2} \right\},$$

(7.17)

and from (7.8) and (7.16)

$$\frac{1}{T_2^I} = S(S+1)\left\{ \tfrac{4}{3}\delta + \tfrac{1}{3}A^2\tau_e \left[1 + \frac{1}{1+(\omega_I-\omega_S)^2\tau_e^2} \right] \right\}$$

$$\frac{1}{T_2^S} = I(I+1)\left\{ \tfrac{4}{3}\delta + \tfrac{1}{3}A^2\tau_e \left[1 + \frac{1}{1+(\omega_I-\omega_S)^2\tau_e^2} \right] \right\}.$$

(7.18)

For the special case of $I=S=\tfrac{1}{2}$, we find from (7.8), (7.14), (7.17) and (7.18) that

$$\rho = \rho' = \delta + \tfrac{1}{2}A^2 \frac{\tau_e}{1+(\omega_I-\omega_S)^2\tau_e^2}$$

$$\sigma = \sigma' = \tfrac{1}{2}\delta - \tfrac{1}{2}A^2 \frac{\tau_e}{1+(\omega_I-\omega_S)^2\tau_e^2}$$

(7.19)

$$\frac{1}{T_2^I} = \frac{1}{T_2^S} = \frac{1}{T_2} = \delta + \tfrac{1}{4}A^2\tau_e \left[1 + \frac{1}{1+(\omega_I-\omega_S)^2\tau_e^2} \right].$$

We can solve these equations for δ, τ_e and A, which will be expressed in terms of ρ, σ and T_2. We see from (7.19) that, if $\sigma < 0$, the scalar coupling is more important than the dipolar coupling for the relaxation processes.

7.3 Scalar relaxation of the second kind

We now turn our attention to the scalar interaction where A is a constant and S is time dependent. We consider the relaxation effect of the interaction Hamiltonian $\hbar A(\mathbf{I}\cdot\mathbf{S}(t))$, where S has for some different interaction a relaxation rate which is much faster than both the angular frequency A and the chemical exchange rate τ_e^{-1}. The different interaction might for $S > \tfrac{1}{2}$ be the quadrupole interaction of the nucleus with an electric field. The problem is treated by combining the spin S with the lattice and assuming that on account of its fast relaxation rate the spin S is in thermal equilibrium with the lattice.

Let us therefore write

$$\hbar G(t) = \hbar A(\mathbf{S}(t)\cdot\mathbf{I}),$$

(7.20)

since we can permute the order of the two independent spin vectors. Putting

$$H_q(t) = AS_q(t) \tag{7.21}$$

we express (7.20) as

$$\hbar G(t) = \sum_{q=-1}^{1} (-)^q H_{-q}(t) I_q.$$

Our classical discussion of $H_q(t)$ in sections 3.3 and 3.4 was based on the supposition that $H_q(t)$ is a commuting quantity. This is no longer true for $H_q(t)$ defined in (7.21) and a quantum mechanical treatment of the problem must be employed. A proper exposition of this would be lengthy (Abragam, 1961, pp. 283–9, 310) and, since we shall not require it for any other type of relaxation mechanism, we shall omit it. The most important result for experiment is that the relaxation rates of the spin I nucleus are given by

$$\frac{1}{T_1} = \tfrac{2}{3}A^2 S(S+1)\frac{\tau_T}{1+(\omega_I-\omega_S)^2\tau_T^2}$$

$$\frac{1}{T_2} = \tfrac{1}{3}A^2 S(S+1)\left\{\tau_L + \frac{\tau_T}{1+(\omega_I-\omega_S)^2\tau_T^2}\right\}. \tag{7.22}$$

In these equations τ_L, τ_T are the longitudinal, transverse relaxation times for the spin S, when its coupling with the spin I is neglected. As in the preceding section, there is no interference between the dipolar and scalar interactions.

8

Relaxation by chemical shift

8.1 Nuclear shielding

The general theory of nuclear shielding for an isolated molecule was derived by Ramsey (1950). For purposes of comparison it is convenient to present results in the notation of Abragam (1961, pp. 173 *et seq.*). Considering the case of liquids we fix our attention on a nucleus with magnetic moment μ in a rotating molecule whose ground state is described by a wave function $|o\lambda\rangle$, where λ specifies the rotational state and o the other degrees of freedom. An external uniform magnetic field $\mathbf{H}$ acts in a fixed direction. We take a coordinate system S with origin at the nucleus and axes fixed with reference to the laboratory. The position of the kth electron in the molecule is denoted by $\mathbf{r}_k$, and its mass and charge are denoted by m and e. The contribution O_2 to the energy arising from the coupling of the moment μ of the nucleus with the magnetic field of the currents induced by the Larmor precession of the electrons in the field is given by

$$O_2 = \frac{e^2}{2mc^2} \sum_{k=1}^{n} ((\boldsymbol{\mu} \times \mathbf{r}_k) \cdot (\mathbf{H} \times \mathbf{r}_k)) r_k^{-3} \tag{8.1}$$

summed over all the electrons of the molecule. Now in general for three vectors $\mathbf{A}(A_x, A_y, A_z)$, $\mathbf{B}(B_x, B_y, B_z)$, $\mathbf{r}(x, y, z)$ we have that

$$
\begin{aligned}
&((\mathbf{A} \times \mathbf{r}) \cdot (\mathbf{B} \times \mathbf{r})) \\
&= (A_y z - A_z y)(B_y z - B_z y) + (A_z x - A_x z)(B_z x - B_x z) \\
&\quad + (A_x y - A_y x)(B_x y - B_y x) \\
&= A_x B_x (y^2 + z^2) + A_y B_y (z^2 + x^2) + A_z B_z (x^2 + y^2) \\
&\quad - (A_y B_z + A_z B_y) yz - (A_z B_x + A_x B_z) zx - (A_x B_y + A_y B_x) xy \\
&= \mathbf{A} \cdot V \cdot \mathbf{B}, \tag{8.2}
\end{aligned}
$$

where

$$V = \tfrac{2}{3}r^2 \begin{bmatrix} 1 & 0 & 0 \\ 0 & 1 & 0 \\ 0 & 0 & 1 \end{bmatrix}$$

$$+ \begin{bmatrix} y^2 + z^2 - 2r^2/3 & -xy & -xz \\ -yx & z^2 + x^2 - 2r^2/3 & -yz \\ -zx & -zy & x^2 + y^2 - 2r^2/3 \end{bmatrix}, \quad (8.3)$$

the sum of a multiple $V^{(1)}$ of the unit matrix and a symmetric traceless matrix $V^{(2)}$. Thus, from (8.1)–(8.3),

$$O_2 = \frac{e^2}{2mc^2} \sum_{k=1}^{n} \{(\boldsymbol{\mu} \cdot V_k^{(1)} \cdot \mathbf{H}) + (\boldsymbol{\mu} \cdot V_k^{(2)} \cdot \mathbf{H})\} r_k^{-3}, \quad (8.4)$$

where $V_k^{(1)}$, $V_k^{(2)}$ are obtained by making the substitutions

$$r \mapsto r_k, \qquad x \mapsto r_{k1}, \qquad y \mapsto r_{k2}, \qquad z \mapsto r_{k3}$$

in the expressions for $V^{(1)}$, $V^{(2)}$ as given by (8.3).

From section 3.1 and from (A.10) and (A.12) of Appendix A we see that the expectation value of O_2 for the state described by $|o\lambda)$ may be written $(o\lambda | O_2 | o\lambda)$. Then from (8.3) and (8.4)

$$(o\lambda | O_2 | o\lambda) = (\boldsymbol{\mu} \cdot T^{(1)} \cdot \mathbf{H}) + (\boldsymbol{\mu} \cdot T^{(2)} \cdot \mathbf{H}), \quad (8.5)$$

where

$$T_{pq}^{(1)} = \frac{e^2}{3mc^2} \delta_{pq} \left(o\lambda \left| \sum_{k=1}^{n} r_k^{-1} \right| o\lambda \right) \quad (8.6)$$

$$T_{pq}^{(2)} = \frac{e^2}{2mc^2} \sum_{k=1}^{n} \left(o\lambda \left| \frac{\delta_{pq}}{3r_k} - \frac{r_{kp} r_{kq}}{r_k^3} \right| o\lambda \right). \quad (8.7)$$

When applying (8.5)–(8.7) to a liquid we use the knowledge that the thermal motion of the environment compels the molecule to continually change its rotational state. Since the average over these states of r_{kp}^2 is $r_k^2/3$ and since the average of $r_{kp} r_{kq}$ vanishes for $q \neq p$, the average of $T_{pq}^{(2)}$ vanishes. Hence $T_{pq}^{(2)}$ does not contribute to nuclear shielding. On the other hand, adding the term $(\boldsymbol{\mu} \cdot T^{(1)} \cdot \mathbf{H})$ of (8.5) to the unperturbed energy $-(\boldsymbol{\mu} \cdot \mathbf{H})$ we obtain

$$-\sum_{pq} \mu_p (\delta_{pq} - T_{pq}^{(1)}) H_q,$$

that is

$$-\sum_{p} \mu_p (1 - T_{pp}^{(1)}) H_p.$$

Thus in the expression $-(\boldsymbol{\mu}\cdot\mathbf{H})$ for the interaction energy the field H is altered to $(1-\sigma_{\mathrm{d}})H$, where by (8.6)

$$\sigma_{\mathrm{d}}=\frac{e^2}{3mc^2}\left(o\left|\sum_k r_k^{-1}\right|o\right),\qquad(8.8)$$

the scalar r_k^{-1} being independent of λ. Since σ_{d} is positive, we have a shielding or screening effect on the field H. The subscript d is affixed to σ because the effect is diamagnetic, and σ_{d} is called the *shielding constant*. The shielding effect is known as *chemical shift*, or more precisely *isotropic chemical shift*. Since σ_{d} is independent of H, it produces a constant shift in the energy levels for the interaction of the nucleus with the external field. If to find the order of magnitude of σ_{d} we take $(o|r_k^{-1}|o)$ to be the reciprocal of the Bohr radius $h^2/(me^2)$, it will follow from (8.8) that

$$\sigma_{\mathrm{d}}\approx\frac{e^2}{3mc^2}\frac{me^2}{h^2}=\frac{1}{3}\left(\frac{e^2}{hc}\right)^2=\frac{1}{3\times137^2},\qquad(8.9)$$

which is of order 10^{-5}.

Starting with a constant magnetic field in the z-direction of the laboratory coordinate system S we denote by $\mathbf{H}_0$ this constant field multiplied by $1-\sigma_{\mathrm{d}}$. Let us consider a nucleus with spin operator $\mathbf{I}$ and gyromagnetic ratio γ, so that its magnetic moment $\boldsymbol{\mu}$ is $\gamma h\mathbf{I}$ and the energy of the Zeeman coupling to $\mathbf{H}_0$ is $-\gamma h(\mathbf{I}\cdot\mathbf{H}_0)$. We denote this by $h\mathcal{H}_0$. Having accounted for the first term on the right hand side of (8.4) by introducing the screening effect we are left with an interaction Hamiltonian $hG(t)$ given by

$$hG(t)=\frac{e^2}{2mc^2}\sum_{k=1}^{n}(\boldsymbol{\mu}\cdot V_k^{(2)}\cdot\mathbf{H}_0)r_k^{-3}.\qquad(8.10)$$

By reason of the random fluctuations of $\mathbf{r}_k$ the matrix $V_k^{(2)}$ and therefore $G(t)$ are random functions of the time. Moreover on comparing (8.6) with (8.7) and employing (8.9) we deduce that the expectation value of $hG(t)$ is very small compared with that of $h\mathcal{H}_0$. Hence in (8.10) we are entitled to use either $\mathbf{H}_0$ or the initial field $(1-\sigma_{\mathrm{d}})^{-1}\mathbf{H}_0$. Finally we see from (8.3) that the ensemble average of $G(t)$ vanishes. As explained in sections 3.3 and 3.4, the Hamiltonian $hG(t)$ may give rise to a relaxation mechanism. The existence of such a mechanism was first proposed by McConnell & Holm (1956). It is called relaxation by *anisotropic chemical shift* because of the structure of $V^{(2)}$ as contrasted with that of $V^{(1)}$, which provides isotropic chemical shift. The total Hamiltonian $h\mathcal{H}$ is expressed by

$$h\mathcal{H}=h\mathcal{H}_0+hG(t).\qquad(8.11)$$

We put $\mu = \gamma \hbar \mathbf{I}$ into (8.10) and obtain the quantum mechanical equation

$$\hbar G(t) = \frac{e^2 \gamma \hbar}{2mc^2} \sum_{k=1}^{n} (\mathbf{I} \cdot V_k^{(2)} \cdot \mathbf{H}_0) r_k^{-3}. \qquad (8.12)$$

This shows that $\hbar G(t)$ is self-adjoint. On account of its physical significance O_2 is invariant under a rotation of axes. The same is true for the $V^{(1)}$-part of O_2 and therefore for $\hbar G(t)$. If then we write from (3.38)

$$\hbar G(t) = \hbar \sum_{q=-2}^{2} (-)^q H_{-q}(t) A^{(q)}, \qquad (8.13)$$

we deduce from Appendix B that $H_q(t)$ is a component of a spherical tensor of rank 2. Moreover it can only be a function of orientational variables: angular velocities do not enter into our discussion. We therefore deduce from (C.32) that

$$\langle H_q^*(0) H_{q'}(t) \rangle = \frac{1}{5} \delta_{qq'} \sum_{m,m'=-2}^{2} H_m' \langle R^+(t) \rangle_{mm'} H_{m'}'^*, \qquad (8.14)$$

where H_p' is the pth spherical tensor component in a body fixed coordinate frame S'. In (8.14) $R(t)$ is the rotation operator associated with the rotation of S' from its orientation at time zero to its orientation at time t. We saw in subsection 2.4.4 that this is identical with the rotation operator associated with the coordinate frame of the Euler–Langevin equations (2.57).

In order to specify $A^{(q)}$ in (8.13) we proceed as in section 4.1 to construct a $j = 2$ representation from the components of $\mathbf{I}$ and of $\mathbf{H}$, which for the moment we shall not identify with $\mathbf{H}_0$, as follows:

$$A^{(0)} = 3H_z I_z - (\mathbf{H} \cdot \mathbf{I})$$

$$A^{(\pm 1)} = \mp \frac{6^{1/2}}{2} (H_z I_\pm + H_\pm I_z) \qquad (8.15)$$

$$A^{(\pm 2)} = \frac{6^{1/2}}{2} H_\pm I_\pm,$$

where

$$H_\pm = H_x \pm i H_y, \qquad I_\pm = I_x \pm i I_y. \qquad (8.16)$$

Equations (8.15) may be compared with (4.18). Still retaining $\mathbf{H}$ in place of $\mathbf{H}_0$ we express (8.12) as

$$\hbar G(t) = \gamma \hbar \sum_{q=-2}^{2} (\mathbf{I} \cdot \Sigma^{(2)} \cdot \mathbf{H}) \qquad (8.17)$$

with

$$\Sigma^{(2)} = \frac{e^2}{2mc^2} \sum_k V_k^{(2)} r_k^{-3}.$$

We now choose a body frame S'' shown in Fig. 8.1 as that with origin at the nucleus under examination and with axes Ox'', Oy'', Oz'' in directions that will make the real, symmetric and traceless matrix $\Sigma^{(2)}$ diagonal and real (Littlewood, 1950, p. 48). We denote the diagonal elements of the matrix in the coordinate frame S'' by $\delta_{x''}, \delta_{y''}, \delta_{z''}$. S'' and the laboratory system S have a common origin, and so S may be brought to S'' by a rotation. As we see from (B.15), this leads to a unitary transformation of matrices. Since the trace is invariant under a unitary transformation, as was shown in (A.19), we deduce that

$$\delta_{x''} + \delta_{y''} + \delta_{z''} = 0.$$

Fig. 8.1 The coordinate system S' with origin at the centre of mass of the molecule and axes in the directions of the principal axes of inertia, and the coordinate system S'' with origin at the nucleus and axes in the directions that make $\Sigma^{(2)}$ diagonal. The Euler angles α, β, γ determining the orientation of S'' with respect to S' are found by drawing axes through the centre of mass and parallel to the axes of S''.

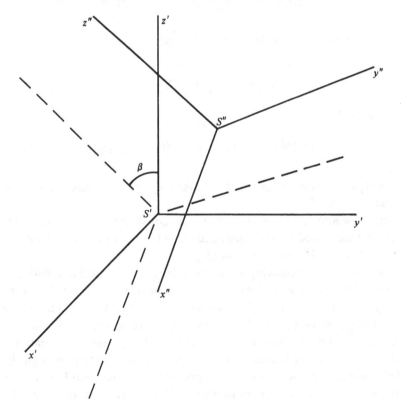

We may therefore express $\delta_{x''}$ and $\delta_{y''}$ in terms of $\delta_{z''}$ and the *asymmetry parameter* ζ by

$$\delta_{x''} = -\tfrac{1}{2}(1-\zeta)\delta_{z''}, \qquad \delta_{y''} = -\tfrac{1}{2}(1+\zeta)\delta_{z''} \qquad (8.18)$$

and so

$$\zeta = \frac{\delta_{x''} - \delta_{y''}}{\delta_{z''}}. \qquad (8.19)$$

Let us now work in the frame S''. Since $\hbar G(t)$ is a rotational invariant, we have from (8.17) and (8.18) that

$$\begin{aligned}
\hbar G(t) &= \gamma\hbar(\mathbf{I}'' \cdot \Sigma^{(2)''} \cdot \mathbf{H}'') \\
&= \gamma\hbar\{ -\tfrac{1}{2}(1-\zeta)\delta_{z''}H_{x''}I_{x''} \\
&\quad -\tfrac{1}{2}(1+\zeta)\delta_{z''}H_{y''}I_{y''} + \delta_{z''}H_{z''}I_{z''}\} \\
&= \tfrac{1}{2}\gamma\hbar\delta_{z''}\{3H_{z''}I_{z''} - (\mathbf{H}'' \cdot \mathbf{I}'') + \tfrac{1}{2}\zeta[H''_+I''_+ + H''_-I''_-]\} \\
&= \tfrac{1}{2}\gamma\hbar\delta_{z''}\left\{ A^{(0)''} + \frac{\zeta}{6^{1/2}}\left[A^{(2)''} + A^{(-2)''}\right]\right\}, \qquad (8.20)
\end{aligned}$$

by (8.15) and (8.16) applied to S''. On comparing (8.20) with

$$\hbar G(t) = \hbar \sum_{q=-2}^{2} (-)^q H''_{-q}A^{(q)''}$$

we deduce that

$$H''_0 = \tfrac{1}{2}\gamma\delta_{z''}, \qquad H''_{\pm 1} = 0, \qquad H''_{\pm 2} = \frac{\zeta\gamma}{2(6)^{1/2}}\delta_{z''}. \qquad (8.21)$$

The H''_q's are time independent because they refer to axes fixed in the rotating molecule.

It is now seen that the reason why we did not identify $\mathbf{H}$ in (8.15) with $\mathbf{H}_0$ was that we wished to apply these equations to the coordinate frame S'', in which $H_{x''}$ and $H_{y''}$ are non-vanishing.

With a view to employing results from Appendix D in the calculation of spectral densities and correlation times we must transform (8.21) to the body frame S' of the Euler–Langevin equations (2.57). As was argued at the beginning of section 6.1 for dipolar interaction, we may regard S' and S'' as having a common origin when we are considering effects of rotational motion only. Suppose that we go from S' to S'' by rotating through the Euler angles α, β, γ as explained in Appendix B, so that from Fig. B.1 β, α are the constant polar, azimuthal angles of the z''-axis with reference to S'.

Then we have from (B.17)

$$H''_{2m} = \sum_{s=-2}^{2} D^2_{sm}(\alpha, \beta, \gamma) H'_{2s},$$ (8.22)

and conversely from (B.13), (B.14), (B.20) and (8.22)

$$H'_{2m} = \sum_{s=-2}^{2} D^{2*}_{ms}(\alpha, \beta, \gamma) H''_{2s}.$$ (8.23)

Writing H'_{2m} as $H_{2m}(\mathbf{a})$, where $\mathbf{a}$ is the unit vector with respect to S' in the direction of the z''-axis, and employing (B.3), (B.4), (B.20) and (B.23) we express the last equation as

$$H_{2m}(\mathbf{a}) = \left(\frac{\pi}{5}\right)^{1/2} \gamma \delta_{z''} Y^*_{2,-m}(\beta, \alpha)$$

$$+ \frac{\zeta \gamma \delta_{z''}}{2(6)^{1/2}} \{D^{2*}_{m2}(\alpha, \beta, \gamma) + D^{2*}_{m,-2}(\alpha, \beta, \gamma)\}.$$ (8.24)

It would obviously be cumbersome to employ (8.24) in (D.11) unless $\sigma(\omega)_{nn'}$ has special symmetry properties. For this reason we shall often make the assumption that the asymmetry parameter vanishes. We see from (8.18) that the z''-axis is then an axis of cylindrical symmetry for the transformed traceless tensor. When $\zeta = 0$, (8.24) reduces to

$$H_{2m}(\mathbf{a}) = \left(\frac{\pi}{5}\right)^{1/2} \gamma \delta_{z''} Y^*_{2,-m}(\beta, \alpha).$$ (8.25)

The components of $H_{2m}(\mathbf{a})$ are expressed explicitly by

$$H_{20}(\mathbf{a}) = \tfrac{1}{4}(3 \cos^2 \beta - 1) \gamma \delta_{z''}$$

$$H_{2,\pm 1}(\mathbf{a}) = \pm (\tfrac{3}{8})^{1/2} \cos \beta \sin \beta \, e^{\pm i\alpha} \gamma \delta_{z''}$$ (8.26)

$$H_{2,\pm 2}(\mathbf{a}) = (\tfrac{3}{32})^{1/2} \sin^2 \beta \, e^{\pm 2i\alpha} \gamma \delta_{z''},$$

as we see from (B.30).

While we have written in (8.13) a generic expression for the interaction Hamiltonian, we have never specified the value of $H_{2q}(t)$ for the laboratory coordinate system. Indeed the use of (8.14) obviated the need of doing so. On substituting (8.25) into (D.15) and employing the properties

$$\tau(\omega)_{nn'} = \tau(\omega)_{n'n} = \tau(\omega)_{-n,-n'},$$

derivable from (D.50)–(D.52), we see that, when the asymmetry parameter vanishes, the correlation time of $H_{2q}(t)$ is τ_2. We have already pointed out that τ_2 is also the correlation time for intramolecular dipolar interaction.

8.2 Relaxation by anisotropic chemical shift

We investigate how nuclear magnetic relaxation results from the interaction Hamiltonian of (8.12), where the constant magnetic field H_0 is in the z-direction of the laboratory coordinate system S (McConnell, 1984a). Equations (8.15) reduce to

$$A^{(0)} = 2H_0 I_z, \qquad A^{(\pm 1)} = \mp \frac{6^{1/2}}{2} H_0 I_\pm, \qquad A^{(\pm 2)} = 0. \qquad (8.27)$$

Consequently (8.13) simplifies to

$$\hbar G(t) = \hbar \sum_{q=-1}^{1} (-)^q H_{-q}(t) A^{(q)}. \qquad (8.28)$$

While non-vanishing elements of $A^{(q)}$ occur only for $q = 0, \pm 1$, it must not be concluded that $A^{(q)}$ are elements of a spherical tensor operator of rank one. Indeed for this to be so $A^{(-1)}, A^{(0)}, A^{(1)}$ would have to be proportional to

$$\frac{I_-}{2^{1/2}}, \qquad I_z, \qquad -\frac{I_+}{2^{1/2}},$$

as we see from (B.29).

We next employ (3.72) to calculate relaxation times. The values of q are, from (8.28), just 0 and ± 1. The reasoning in section 4.2 shows that $q = 0$ is always associated with $j(0)$ and that $q = \pm 1$ is always associated with $j(\omega_0)$. We therefore write down from (3.72)

$$\frac{\mathrm{d}\langle I_r \rangle}{\mathrm{d}t} = \left[\sum_{\alpha\alpha'\beta\beta'} \left\{ j(0) \left[2A^{(0)}_{\alpha\beta} A^{(0)}_{\beta'\alpha'} - \delta_{\alpha\beta} \sum_\sigma A^{(0)}_{\sigma\alpha'} A^{(0)}_{\beta'\sigma} - \delta_{\alpha'\beta'} \sum_\sigma A^{(0)}_{\sigma\beta} A^{(0)}_{\alpha\sigma} \right] \right. \right.$$

$$\left. \left. + j(\omega_0) \sum_{q=1,-1} \left[2A^{(q)}_{\alpha\beta} A^{(q)+}_{\beta'\alpha'} - \delta_{\alpha\beta} \sum_\sigma A^{(q)}_{\sigma\alpha'} A^{(q)+}_{\beta'\sigma} - \delta_{\alpha'\beta'} \sum_\sigma A^{(q)}_{\sigma\beta} A^{(q)+}_{\alpha\sigma} \right] \right\} \right]$$

$$\times \rho(t)_{\beta\beta'} (I_r)_{\alpha'\alpha}. \qquad (8.29)$$

The calculations are simpler than those for dipolar coupling. From (8.27) the coefficient of $j(0)$ in the right hand side of (8.29),

$$\mathrm{tr}\{2\rho(t) A^{(0)} I_r A^{(0)} - \rho(t) I_r A^{(0)2} - \rho(t) A^{(0)2} I_r\}$$

$$= 4H_0^2 \langle 2I_z I_r I_z - I_r I_z^2 - I_z^2 I_r \rangle$$

$$= 4H_0^2 \langle [I_z, [I_r, I_z]] \rangle.$$

When $I_r \equiv I_z$, the double commutator vanishes and the coefficient of $j(0)$ is zero, as could have been expected from our remark after (4.59). When $I_r \equiv I_x$,

$$[I_z, [I_r, I_z]] = -\mathrm{i}[I_z, I_y] = -I_x$$

and the coefficient of $j(0)$ is $-4H_0^2 \langle I_x \rangle$. In the same way it is found that the coefficient of $j(\omega_0)$ in the right hand side of (8.29) is

$$\tfrac{3}{2} H_0^2 \langle [I_-, [I_r, I_+]] + [I_+, [I_r, I_-]] \rangle.$$

It is easily seen by using (4.33) and (4.36) that this expression is equal to $-6H_0^2\langle I_z\rangle$ for $I_r \equiv I_z$ and to $-3H_0^2\langle I_x\rangle$ for $I_r \equiv I_x$.

In order that $\langle I_z\rangle$ should relax to $\langle I_z\rangle_0$ for a steady state we replace $\langle I_z\rangle$ by $\langle I_z\rangle - \langle I_z\rangle_0$ in the equation deduced from (8.29) for $\langle I_z\rangle$. We then obtain

$$\frac{d(\langle I_z\rangle - \langle I_z\rangle_0)}{dt} = -6H_0^2 j(\omega_0)(\langle I_z\rangle - \langle I_z\rangle_0)$$

$$\frac{d\langle I_x\rangle}{dt} = -H_0^2(4j(0) + 3j(\omega_0))\langle I_x\rangle.$$

Thus for anisotropic chemical shift we have a longitudinal relaxation time T_1 given by

$$\frac{1}{T_1} = 6H_0^2 j(\omega_0) \tag{8.30}$$

and a transverse relaxation time T_2, where

$$\frac{1}{T_2} = H_0^2(4j(0) + 3j(\omega_0)). \tag{8.31}$$

We see that in the extreme narrowing approximation when $j(\omega_0)$ is replaced by $j(0)$

$$\frac{1}{T_1} = \frac{6}{7T_2} = 6H_0^2 j(0). \tag{8.32}$$

8.3 Relaxation times for molecular models

We return to (3.48) in the laboratory system, which we write more explicitly as

$$j(\omega) = \frac{1}{2} \int_{-\infty}^{\infty} \langle H_0(\mathbf{a}(0)) H_0(\mathbf{a}(t)) \rangle\, e^{-i\omega t}\, dt \tag{8.33}$$

to indicate that H_q depends on a direction fixed with respect to a molecular coordinate frame and specified at time t by a unit vector $\mathbf{a}(t)$ with respect to the laboratory frame S. We take the molecular frame to be S'; this is the frame to which the Euler–Langevin equations (2.57) are referred. We deduce from (8.33) and (D.11) that

$$j(\omega) = \frac{1}{10} \sum_{n,n' = -2}^{2} H_n' H_{n'}'^* \sigma(\omega)_{nn'}. \tag{8.34}$$

The value of H_m' is to be found from (8.21) and (8.23). The quantity $\sigma(\omega)$ is independent of the relaxation mechanism. For a rigid molecule without any special symmetry, the value of $\sigma(\omega)_{nn'}$ may be obtained from (D.53).

At this stage it is convenient to distinguish between a spherical and a non-spherical molecular model. For the first model we see from (D.24) that $\sigma(\omega)_{nn'} = \delta_{nn'}\sigma(\omega)_{00}$. Moreover on account of the spherical symmetry we may identify S' with the coordinate system S'' of section 8.1 and thence (8.34) becomes

$$j(\omega) = \tfrac{1}{10}\sigma(\omega)_{00} \sum_{n=-2}^{2} |H_n''|^2,$$

that is

$$j(\omega) = \tfrac{1}{40}\gamma^2(1+\tfrac{1}{3}\zeta^2)\delta_{z''}^2\sigma(\omega)_{00}, \qquad (8.35)$$

by (8.21).

The relaxation rates may be found from (8.30)–(8.32), (8.35) and (D.24):

$$\frac{1}{T_1} = \tfrac{3}{10}\gamma^2 H_0^2(1+\tfrac{1}{3}\zeta^2)\delta_{z''}^2\left\{\frac{G'}{B(G'^2+\omega_0'^2)} + \frac{6\kappa}{B}\frac{G'(1+G')-\omega_0'^2}{(G'^2+\omega_0'^2)[(1+G')^2+\omega_0'^2]} + \cdots\right\}$$

$$\frac{1}{T_2} = \tfrac{1}{5}\gamma^2 H_0^2(1+\tfrac{1}{3}\gamma^2)\delta_{z''}^2\left\{\frac{1}{BG'} + \frac{6\kappa}{BG'(1+G')} + \frac{3G'}{4B(G'^2+\omega_0'^2)}\right.$$

$$\left. + \frac{9\kappa}{2B}\frac{G'(1+G')-\omega_0'^2}{(G'^2+\omega_0'^2)[(1+G')^2+\omega_0'^2]} + \cdots\right\}. \qquad (8.36)$$

These show that the chemical shift relaxation rates have a strong dependence on the external field H_0, which was not present in (6.4) for dipolar relaxation. For the extreme narrowing approximation we now find that

$$\frac{1}{T_1} = \frac{6}{7T_2} = \tfrac{1}{20}\gamma^2 H_0^2(1+\tfrac{1}{3}\zeta^2)\delta_{z''}^2(1+\tfrac{11}{2}\kappa - \tfrac{83}{6}\kappa^2 + \cdots)\frac{I_1 B}{kT}. \qquad (8.37)$$

The correlation time τ_2 is given by (6.5).

In the rotational diffusion limit we have from (8.35) and (D.30)

$$j(\omega) = \tfrac{1}{120}\gamma^2(1+\tfrac{1}{3}\zeta^2)\delta_{z''}^2\frac{I_1 B}{kT\left[1+\left(\dfrac{I_1 B\omega}{6kT}\right)^2\right]}. \qquad (8.38)$$

From this it follows that

$$\frac{1}{T_1} = \tfrac{1}{20}\gamma^2 H_0^2(1+\tfrac{1}{3}\zeta^2)\delta_{z''}^2\frac{I_1 B}{kT\left[1+\left(\dfrac{I_1 B\omega_0}{6kT}\right)^2\right]} \qquad (8.39)$$

$$\frac{1}{T_2} = \tfrac{1}{30}\gamma^2 H_0^2(1+\tfrac{1}{3}\zeta^2)\delta_{z''}^2\frac{I_1 B}{kT}\left\{1+\frac{3/4}{1+\left(\dfrac{I_1 B\omega_0}{6kT}\right)^2}\right\}$$

and in the extreme narrowing approximation

$$\frac{1}{T_1}=\frac{6}{7T_2}=\tfrac{1}{20}\gamma^2 H_0^2(1+\tfrac{1}{3}\zeta^2)\delta_{z''}^2\frac{I_1 B}{kT}.\qquad(8.40)$$

The correlation time τ_2 is now given by (D.31) and we immediately verify from (8.21) and (8.38) that (D.18) is satisfied. Replacing $\gamma^2 H_0^2$ by ω_0^2 in (8.39) and comparing these with (6.11) and (6.12) we see that in the Debye approximation T_1 and T_2 are decreasing functions of Larmor frequency for anisotropic chemical shift, whereas they are increasing functions for dipolar interactions.

Turning to the case of non-spherical molecular models we reduce the complexity of the calculations by making the assumption that the asymmetry parameter ζ vanishes. This allows us to take the values of H'_m from (8.25) and on inserting them into (8.34) we obtain

$$j(\omega)=\tfrac{1}{50}\pi\gamma^2\delta_{z''}^2\sum_{n,n'=-2}^{2} Y_{2,-n}(\beta,\alpha)Y^*_{2,-n'}(\beta,\alpha)\sigma(\omega)_{nn'}.$$

From inspection of (D.53) we see that $\sigma(\omega)_{nn'}=\sigma(\omega)_{-n,-n'}$, and so

$$j(\omega)=\tfrac{1}{50}\pi\gamma^2\delta_{z''}^2\sum_{n,n'=-2}^{2} Y_{2n}(\beta,\alpha)Y^*_{2n'}(\beta,\alpha)\sigma(\omega)_{nn'}.\qquad(8.41)$$

On the other hand we have from (6.2) that for intramolecular dipolar relaxation of like spins

$$j(\omega)=\frac{3\pi\gamma^4\hbar^2}{25r^6}\sum_{n,n'=-2}^{2} Y_{2n}(\theta',\phi')Y^*_{2n'}(\theta',\phi')\sigma(\omega)_{nn'}.\qquad(8.42)$$

Comparing (8.41) with (8.42) we see that for any molecular model the value of $j(\omega)$ for relaxation by anisotropic chemical shift may be obtained from the corresponding value for relaxation by dipolar interaction by making the substitutions

$$\theta'\mapsto\beta,\qquad \phi'\mapsto\alpha,\qquad r^{-6}\mapsto\tfrac{1}{6}\delta_{z''}^2\gamma^{-2}\hbar^{-2}.\qquad(8.43)$$

When this has been done, we deduce T_1 and T_2 from (8.30)–(8.32).

The replacements (8.43) enable us to find $j(\omega)$ not only in inertial theory but also in rotational diffusion theory for the different molecular models, for finite ω and for extreme narrowing. Thus for an asymmetric molecule in inertial theory we deduce from (6.23) and (8.43) that the spectral density for anisotropic chemical shift

$$j(\omega)=\frac{\gamma^2\delta_{z''}^2}{160}\{3\sin^4\beta(A+A^*)+3\sin^2 2\beta(B+B^*)$$

$$+(3\cos^2\beta-1)^2(C+C^*)+2(6)^{1/2}(3\cos^2\beta-1)\sin^2\beta\cos 2\alpha(D+D^*)$$

$$-3\sin^2 2\beta\cos 2\alpha(E+E^*)+3\sin^4\beta\cos 4\alpha(F+F^*)\},\qquad(8.44)$$

where A, B, C, D, E, F are defined in (D.51) and (D.52). Referring to the last paragraph of section 8.1 we see that the correlation time τ_2' is obtained by making the replacements $\theta' \mapsto \beta$, $\phi' \mapsto \alpha$ in the right hand side of (6.26).

For the symmetric rotator we find from (6.42) and (8.43) that

$$j(\omega) = \frac{3\gamma^2 \delta_{z''}^2}{40} \left\{ \frac{\sin^4 \beta (D_1 + 2D_3)\left(1 + \dfrac{2D_1}{B_1} + \dfrac{4D_3}{B_3}\right)}{(2D_1 + 4D_3)^2 + \omega^2} \right.$$

$$+ \frac{2\sin^2 \beta \cos^2 \beta (5D_1 + D_3)\left(1 + \dfrac{5D_1}{B_1} + \dfrac{D_3}{B_3}\right)}{(5D_1 + D_3)^2 + \omega^2}$$

$$\left. + \frac{(3\cos^2 \beta - 1)^2 D_1 \left(1 + \dfrac{6D_1}{B_1}\right)}{36D_1^2 + \omega^2} \right\}. \qquad (8.45)$$

From (6.47) we likewise obtain in rotational diffusion theory

$$j(\omega) = \frac{3\gamma^2 \delta_{z''}^2}{40} \left\{ \frac{\sin^4 \beta (D_1 + 2D_3)}{(2D_1 + 4D_3)^2 + \omega^2} + \frac{2\sin^2 \beta \cos^2 \beta (5D_1 + D_3)}{(5D_1 + D_3)^2 + \omega^2} \right.$$

$$\left. + \frac{(3\cos^2 \beta - 1)^2 D_1}{36D_1^2 + \omega^2} \right\}. \qquad (8.46)$$

We may note that for the extreme narrowing approximation in rotational diffusion theory (8.32) and (8.46) yield

$$\frac{1}{T_1} = \frac{6}{7T_2} = \frac{3\gamma^2 H_0^2 \delta_{z''}^2}{40} \left\{ \frac{3\sin^4 \beta}{2D_1 + 4D_3} + \frac{3\sin^2 2\beta}{5D_1 + D_3} + \frac{(3\cos^2 \beta - 1)^2}{6D_1} \right\}. \qquad (8.47)$$

The correlation times τ_2 may be found on replacing θ' by β in (6.45) and (6.48).

When the molecule is linear, we deduce from (8.41) and (D.40)–(D.42) that

$$j(\omega) = \frac{\pi}{50} \gamma^2 \delta_{z''}^2 \{ |Y_{20}(\beta, \alpha)|^2 \sigma(\omega)_{00} + 2|Y_{21}(\beta, \alpha)|^2 \sigma(\omega)_{11}$$

$$+ 2|Y_{22}(\beta, \alpha)|^2 \sigma(\omega)_{22} \},$$

so that, by (B.30),

$$j(\omega) = \frac{\gamma^2 \delta_{z''}^2}{160} \{ (3\cos^2 \beta - 1)^2 \sigma(\omega)_{00} + 3\sin^2 2\beta \sigma(\omega)_{11} + 3\sin^4 \beta \sigma(\omega)_{22} \}.$$

$$(8.48)$$

The values of T_1^{-1}, T_2^{-1} may be obtained from (8.30)–(8.32), (D.40)–(D.42) and (8.48).

To obtain simple expressions for $j(\omega)$ and the relaxation times let us suppose that the z''-axis of the molecular coordinate system S'' coincides with the z'-axis of S in the case of the linear molecule and that in the case of the symmetric rotator molecule the z''-axis is parallel to the z'-axis. Then the angle β vanishes, and in the second case we have from (8.45), (8.30) and (8.31) the inertial results

$$j(\omega) = \frac{3\gamma^2\delta_{z''}^2}{10} \frac{D_1\left(1+\dfrac{6D_1}{B_1}\right)}{36D_1^2+\omega^2} \tag{8.49}$$

$$\frac{1}{T_1} = \frac{9\gamma^2 H_0^2\delta_{z''}^2 D_1\left(1+\dfrac{6D_1}{B_1}\right)}{5(36D_1^2+\omega^2)} \tag{8.50}$$

$$\frac{1}{T_2} = \frac{3\gamma^2 H_0^2\delta_{z''}^2}{10}\left\{\frac{1+\dfrac{6D_1}{B_1}}{9D_1}+\frac{3D_1\left(1+\dfrac{6D_1}{B_1}\right)}{36D_1^2+\omega^2}\right\} \tag{8.51}$$

and in the extreme narrowing approximation

$$\frac{1}{T_1} = \frac{6}{7T_2} = \frac{\gamma^2 H_0^2\delta_{z''}^2\left(1+\dfrac{6D_1}{B_1}\right)}{20D_1}. \tag{8.52}$$

The results for the rotational diffusion theory obtained by putting $D_1 = kT/(I_1 B_1)$ and $D_1/B_1 = 0$ in (8.49)–(8.52) are

$$j(\omega) = \frac{\gamma^2\delta_{z''}^2 I_1 B_1}{120kT}\frac{1}{1+\left(\dfrac{I_1 B_1\omega}{6kT}\right)^2} \tag{8.53}$$

$$\frac{1}{T_1} = \frac{\gamma^2 H_0^2\delta_{z''}^2 I_1 B_1}{20kT}\frac{1}{1+\left(\dfrac{I_1 B_1\omega_0}{6kT}\right)^2} \tag{8.54}$$

$$\frac{1}{T_2} = \frac{\gamma^2 H_0^2\delta_{z''}^2 I_1 B_1}{30kT}\left\{1+\frac{3/4}{1+\left(\dfrac{I_1 B_1\omega_0}{6kT}\right)^2}\right\} \tag{8.55}$$

$$\frac{1}{T_1} = \frac{6}{7T_2} = \frac{\gamma^2 H_0^2\delta_{z''}^2 I_1 B_1}{20 kT}. \tag{8.56}$$

Equations (8.53)–(8.56) become identical with (8.38)–(8.40) for the spherical molecules when the asymmetry parameter ζ vanishes and B_1 is identified with B. In fact this result follows from our discussions at the ends of Appendix D and of section 6.5 (McConnell, 1986b).

When $\beta=0$ for the linear molecule (8.48) yields

$$j(\omega) = \tfrac{1}{40}\gamma^2\delta_{z''}^2\sigma(\omega)_{00},\tag{8.57}$$

where $\sigma(\omega)_{00}$ is now given by (D.40). In the rotational diffusion limit the value of $\sigma(\omega)_{00}$ comes from (D.46), which is identical with (D.30) for the sphere. We then obtain T_1 and T_2 by putting $\zeta=0$ in (8.39) and (8.40), and so in the extreme narrowing approximation

$$\frac{1}{T_1} = \frac{6}{7T_2} = \frac{\gamma^2 H_0^2 \delta_{z''}^2}{20}\frac{I_1 B}{kT}.\tag{8.58}$$

To compare this with the results of Spiess *et al.* (1971) let us write

$$\delta_{x''} = \delta_{y''} = \sigma_\perp, \qquad \delta_{z''} = \sigma_\parallel$$

and define the *shielding anisotropy* $\Delta\sigma$ by

$$\Delta\sigma = \sigma_\parallel - \sigma_\perp.\tag{8.59}$$

Then, from (8.18) with $\zeta=0$,

$$\Delta\sigma = \tfrac{3}{2}\delta_{z''}\tag{8.60}$$

and, from (8.58), (8.59) and (8.60),

$$\frac{1}{T_1} = \frac{I_1\omega_0^2(\Delta\sigma)^2 B}{45kT}.\tag{8.61}$$

Now the components of angular velocity of the linear rotator satisfy (McConnell, 1980*b*, section 10.2)

$$\langle\omega_i(t_1)\omega_j(t_2)\rangle = \delta_{ij}\frac{kT}{I_1}e^{-B|t_1-t_2|}.$$

From this and (2.24) we deduce that the correlation time for a component of angular velocity is B^{-1}. In the notation of Spiess *et al.* this is denoted by τ_{SR} and we may then identify their equation [8] with (8.61).

9

Relaxation by quadrupole interaction

9.1 Quadrupole interactions

We consider the electrostatic interaction between the protons in an atomic nucleus and an external electron. We take a laboratory coordinate system S through the centre of mass of the nucleus. Let $\mathbf{r}_p$ be the position vector of the proton, $\mathbf{r}_e$ the position vector of the electron and χ the angle between $\mathbf{r}_e$ and $\mathbf{r}_p$. Then expanding the reciprocal of the distance between the proton and electron as a series in Legendre polynomials we see from Fig. 9.1 that

$$\frac{1}{|\mathbf{r}_e - \mathbf{r}_p|} = \sum_{l=0}^{\infty} \frac{r_p^l}{r_e^{l+1}} P_l(\cos \chi). \tag{9.1}$$

On employing the addition theorem (Rose, 1957, p. 60)

$$P_l(\cos \chi) = \frac{4\pi}{2l+1} \sum_{m=-l}^{l} Y_{lm}^*(\theta_p, \phi_p) Y_{lm}(\theta_e, \phi_e),$$

where (θ_p, ϕ_p), (θ_e, ϕ_e) are the spherical polar angles of the proton and electron, respectively, we deduce from (9.1) that

$$\frac{1}{|\mathbf{r}_e - \mathbf{r}_p|} = \sum_{l=0}^{\infty} \frac{4\pi r_p^l}{(2l+1)r_e^{l+1}} \sum_{m=-l}^{l} Y_{lm}^*(\theta_p, \phi_p) Y_{lm}(\theta_e, \phi_e).$$

If e denotes the charge on the proton and Z is the atomic number of the nucleus, the electrostatic energy for the interaction between the electron and the nucleus

$$W = -\sum_{p=1}^{Z} \sum_{l=0}^{\infty} \sum_{m=-l}^{l} \frac{4\pi e^2 r_p^l}{(2l+1)r_e^{l+1}} Y_{lm}^*(\theta_p, \phi_p) Y_{lm}(\theta_e, \phi_e). \tag{9.2}$$

We may, by (B.3), express (9.2) as

$$W = \sum_{p=1}^{Z} W_p, \qquad W_p = \sum_{l=0}^{\infty} \sum_{q=-l}^{l} (-)^q F_{l,-q} A^{lq}, \tag{9.3}$$

where

$$F_{lq} = -e\left(\frac{4\pi}{2l+1}\right)^{1/2} r_e^{-(l+1)} Y_{lq}(\theta_e, \phi_e) \tag{9.4}$$

$$A^{lq} = e \left(\frac{4\pi}{2l+1} \right)^{1/2} r_p^l Y_{lq}(\theta_p, \phi_p).$$ (9.5)

We see that F_{lq} and A^{lq} are both components of spherical tensors of rank l.

To discuss A^{lq} we introduce quantum theory describing the nuclear state by a wave function $\psi(\mathbf{r}_{p_1}, \mathbf{r}_{p_2}, \ldots, \mathbf{r}_{p_A})$, where A is the mass number. The wave function is unaltered or is multiplied by -1 for an inversion through the origin (Ramsey, 1953, p. 23). To invert A^{lq} we make the substitutions

$$\theta_p \mapsto \pi - \theta_p, \qquad \phi_p \mapsto \phi_p + \pi$$

and we find from (B.1) that $A^{lq} \mapsto (-)^l A^{lq}$. To obtain the mean value of A^{lq} we integrate $\psi^* A^{lq} \psi$ over the coordinates, and we see that the integral vanishes when l is odd.

We therefore investigate even values of l. When $l=0$, then from (B.1), (9.4) and (9.5)

$$F_{00} = -e(4\pi)^{1/2} r_e^{-1} Y_{00}(\theta_e, \phi_e) = -e r_e^{-1}, \qquad A^{00} = e,$$

and the contribution to W_p is $-e^2/r_e$. This is the usual interaction energy between opposite charges in the limit of $r_p \to 0$.

For $l=2$, (9.4) and (9.5) give

$$F_{2q} = -e \left(\frac{4\pi}{5} \right)^{1/2} r_e^{-3} Y_{2q}(\theta_e, \phi_e)$$ (9.6)

$$A^{2q} = e \left(\frac{4\pi}{5} \right)^{1/2} r_p^2 Y_{2q}(\theta_p, \phi_p).$$ (9.7)

To understand the physical significance of F_{2q}, let us consider the electrostatic field produced at a point $\mathbf{r}(x, y, z)$ by the electron at $\mathbf{r}_e$. Writing $|\mathbf{r} - \mathbf{r}_e| = R$ we have the electrostatic potential $U = -e/R$. From

$$R^2 = (x - x_e)^2 + (y - y_e)^2 + (z - z_e)^2$$

Fig. 9.1 The position vectors $\mathbf{r}_p$ of a proton in the nucleus and $\mathbf{r}_e$ of an electron outside.

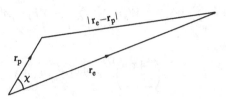

we deduce that

$$\frac{\partial U}{\partial x} = \frac{dU}{dR}\frac{\partial R}{\partial x} = \frac{e(x-x_e)}{R^3}$$

$$\frac{\partial U}{\partial y} = \frac{e(y-y_e)}{R^3}, \qquad \frac{\partial U}{\partial z} = \frac{e(z-z_e)}{R^3}$$

$$\frac{\partial^2 U}{\partial x^2} = \frac{e}{R^3} - \frac{3e(x-x_e)^2}{R^5}, \qquad \frac{\partial^2 U}{\partial y^2} = \frac{e}{R^3} - \frac{3e(y-y_e)^2}{R^5}$$

$$\frac{\partial^2 U}{\partial z^2} = \frac{e}{R^3} - \frac{3e(z-z_e)^2}{R^5}, \qquad \frac{\partial^2 U}{\partial x\,\partial y} = -\frac{3e(x-x_e)(y-y_e)}{R^5}$$

$$\frac{\partial^2 U}{\partial x\,\partial z} = -\frac{3e(x-x_e)(z-z_e)}{R^5}, \qquad \frac{\partial^2 U}{\partial y\,\partial z} = -\frac{3e(y-y_e)(z-z_e)}{R^5}.$$

If we evaluate these quantities at the origin putting $x=y=z=0$ and express x_e, y_e, z_e in terms of the spherical polar coordinates, we obtain $R=r_e$ and

$$\frac{\partial^2 U}{\partial x^2} = \frac{e(1-3\sin^2\theta_e\cos^2\phi_e)}{r_e^3}, \qquad \frac{\partial^2 U}{\partial y^2} = \frac{e(1-3\sin^2\theta_e\sin^2\phi_e)}{r_e^3}$$

$$\frac{\partial^2 U}{\partial z^2} = \frac{e(1-3\cos^2\theta_e)}{r_e^3}, \qquad \frac{\partial^2 U}{\partial x\,\partial y} = -\frac{3e\sin^2\theta_e\cos\phi_e\sin\phi_e}{r_e^3} \qquad (9.8)$$

$$\frac{\partial^2 U}{\partial x\,\partial z} = -\frac{3e\sin\theta_e\cos\theta_e\cos\phi_e}{r_e^3}, \qquad \frac{\partial^2 U}{\partial y\,\partial z} = -\frac{3e\sin\theta_e\cos\theta_e\sin\phi_e}{r_e^3}.$$

Then substituting from (B.30) into (9.6) we deduce from (9.8) that

$$F_{20} = \frac{1}{2}\frac{\partial^2 U}{\partial z^2}, \qquad F_{2,\pm1} = \mp\frac{1}{6^{1/2}}\left(\frac{\partial^2 U}{\partial x\,\partial z} \pm i\frac{\partial^2 U}{\partial y\,\partial z}\right)$$

$$F_{2,\pm2} = \frac{1}{2(6)^{1/2}}\left(\frac{\partial^2 U}{\partial x^2} - \frac{\partial^2 U}{\partial y^2} \pm 2i\frac{\partial^2 U}{\partial x\,\partial y}\right), \qquad (9.9)$$

where

$$U = -\frac{e}{r_e} \qquad (9.10)$$

and the derivatives are calculated at the centre of mass of the nucleus.

We now turn our attention to the quantity $Q_q^{(2)}$ defined by

$$Q_q^{(2)} = \sum_{p=1}^{z} A^{2q} \qquad (9.11)$$

with A^{2q} given by (9.7). It is customary (Abragam, 1961, p. 162) to define the *electric quadrupole moment* eQ of an atomic nucleus in a state with spin angular momentum quantum number I as the expectation value of

$e\sum_{p=1}^{Z}(3z_p^2-r_p^2)$ for the state with $I_z=I$: thus, as was explained in section 8.1, we write

$$Q=\left(II\left|\sum_{p=1}^{Z}(3z_p^2-r_p^2)\right|II\right).\tag{9.12}$$

Then from (B.30), (9.11) and (9.12)

$$Q=2\left(\frac{4\pi}{5}\right)^{1/2}\sum_{p=1}^{Z}r_p^2(II|Y_{20}(\theta_p,\phi_p)|II)$$

$$=\frac{2}{e}(II|Q_0^{(2)}|II),$$

and so

$$(II|Q_0^{(2)}|II)=\tfrac{1}{2}eQ.\tag{9.13}$$

Since by the Wigner–Eckart theorem (C.26)

$$(II|Q_0^{(2)}|II)=C(I2I;\,I0I)(I\|\mathbf{Q}^{(2)}\|I)$$

and since the Clebsch–Gordan coefficient C vanishes unless I,I and 2 can be the lengths of sides of a triangle, it follows from (9.13) that the quadrupole moment is zero unless $I\geqslant 1$.

Let us now find expressions for the components of $Q_q^{(2)}$. From the Wigner–Eckart theorem

$$(Im'|Q_l^{(2)}|Im)=C(I2I;\,mlm')(I\|\mathbf{Q}^{(2)}\|I),\tag{9.14}$$

where $(I\|\mathbf{Q}^{(2)}\|I)$ is the reduced matrix element. Similarly for the operator $V_q^{(2)}$ defined in (C.19)–(C.21)

$$(Im'|V_l^{(2)}|Im)=C(I2I;\,mlm')(I\|\mathbf{V}^{(2)}\|I),$$

which when combined with (9.14) yields

$$\frac{(Im'|Q_l^{(2)}|Im)}{(Im'|V_l^{(2)}|Im)}=\frac{(I\|\mathbf{Q}^{(2)}\|I)}{(I\|\mathbf{V}^{(2)}\|I)},\tag{9.15}$$

which is independent of m,m' and l. On account of the independence with respect to m and m' the matrix elements of $Q_l^{(2)}$ are proportional to those of $V_l^{(2)}$. We therefore deduce that

$$Q_l^{(2)}=\frac{(I\|\mathbf{Q}^{(2)}\|I)}{(I\|\mathbf{V}^{(2)}\|I)}\,V_l^{(2)}.\tag{9.16}$$

On account of the independence of the right hand side of (9.15) with respect to l we may calculate the right hand side by putting $l=0,\ m'=m=I$ in the left hand side, so that

$$\frac{(I\|\mathbf{Q}^{(2)}\|I)}{(I\|\mathbf{V}^{(2)}\|I)}=\frac{(II|Q_0^{(2)}|II)}{(II|V_0^{(2)}|II)}.\tag{9.17}$$

From (C.25)

$$(II|V_0^{(2)}|II) = 6^{-1/2}(II|3I_z^2 - I(I+1)|II)$$

$$= 6^{-1/2}\{3I^2 - I(I+1)\} = \frac{I(2I-1)}{6^{1/2}}$$

and combining this result with (9.13) we deduce from (9.17) that

$$\frac{(I\|\mathbf{Q}^{(2)}\|I)}{(I\|\mathbf{V}^{(2)}\|I)} = \left(\frac{3}{2}\right)^{1/2} \frac{eQ}{I(2I-1)}.$$

Substituting this into (9.16) and taking the values of $V_1^{(2)}$ from (C.23)–(C.25) we find that

$$Q_0^{(2)} = \frac{eQ}{2I(2I-1)}[3I_z^2 - I(I+1)]$$

$$Q_{\mp 1}^{(2)} = \mp\left(\frac{3}{2}\right)^{1/2} \frac{eQ}{2I(2I-1)}[I_z I_\pm + I_\pm I_z] \qquad (9.18)$$

$$Q_{\pm 2}^{(2)} = \left(\frac{3}{2}\right)^{1/2} \frac{eQ}{2I(2I-1)} I_\pm^2.$$

So far we have considered the field of a single electron outside the nucleus which is described by the electrostatic potential U of (9.10). We now consider the field produced by all the electrons in the molecule that contains the atom whose nucleus is under investigation. In principle electrons in neighbouring molecules could also contribute to the field, but it may be shown that such intermolecular contributions are negligible (Bonera & Rigamonti, 1965b). The only electrons that are likely to come inside the nucleus are s-electrons. Since these have spherically symmetric wave functions, the mean value of $3z_p^2 - r_p^2$ will vanish for them and they will not contribute to Q. Thus we need pay attention only to electrons outside the nucleus. We therefore calculate the expectation value of $-e/r_e$ for each electron and sum it over the electrons so as to obtain the rotationally invariant electrostatic potential which we denote by V. We then replace (9.9) by

$$F_0^{(2)} = \tfrac{1}{2}V_{zz}, \qquad F_{\pm 1}^{(2)} = \mp\frac{1}{6^{1/2}}(V_{xz} \pm iV_{yz}),$$

$$F_{\pm 2}^{(2)} = \frac{1}{2(6)^{1/2}}(V_{xx} - V_{yy} \pm 2iV_{xy}), \qquad (9.19)$$

where the subscripts to the V's denote partial derivatives evaluated at the centre of mass of the nucleus. The $F_k^{(2)}$'s, like the F_{2k}'s, are components of a spherical tensor of rank 2. However, unlike F_{2k}, the $F_k^{(2)}$ is not just a multiple of a spherical harmonic. Combining the sum over electrons with the sum over protons in (9.11) we conclude from (9.3) that the Hamiltonian $\hbar G(t)$ for quadrupole interaction is expressible by

$$\hbar G(t) = \sum_{q=-2}^{2} (-)^q F^{(2)}_{-q} Q^{(2)}_q. \qquad (9.20)$$

When the molecule containing the atomic nucleus is subject to thermal motion, as it will be in our investigations, $F^{(2)}_{-q}$ and consequently the interaction Hamiltonian are time dependent. The spin dependence of the Hamiltonian resides in $Q^{(2)}_q$.

The terms for $l=4$ in the expansion of (9.2) provide a *hexadecapole interaction*. Relative to quadrupolar interaction this will be of order $(r_p/r_e)^2$, that is about 10^{-8}. We shall therefore not consider it further.

To form some idea of the relative importance of quadrupole relaxation we take the case of a deuteron in perdeuterated benzene C_6D_6. Then it has been found (Bonera & Rigamonti, 1965b) that $V_{zz} \approx e\,\text{Å}^{-3}$, which in SI units becomes

$$V_{zz} \approx \frac{10^{30} e}{4\pi\varepsilon_0}\,\text{m}^{-3},$$

where ε_0 is the permittivity of free space. For the deuteron Q is equal to $2.77 \times 10^{-31}\,\text{m}^2$ and the magnetic moment is 0.8574 nuclear magnetons (Andrew, 1969, Appendix 2), the value of the nuclear magneton being $5.05 \times 10^{-27}\,\text{J T}^{-1}$. We see from (9.18)–(9.20) that the perturbing Hamiltonian $\hbar G(t)$ is of order eQV_{zz}. On putting $e = 1.6 \times 10^{-19}\,\text{C}$, $\varepsilon_0 = 8.854 \times 10^{-12}\,\text{F m}^{-1}$ we obtain

$$eQV_{zz} = 6.37 \times 10^{-29}\,\text{J}.$$

Taking the external field to produce a flux density of 1 T we find that the unperturbed Hamiltonian

$$\hbar\mathscr{H}_0 = 0.8574 \times 5.05 \times 10^{-27}\,\text{J} = 4.32 \times 10^{-27}\,\text{J}.$$

Hence $\hbar G(t)$ is about two orders of magnitude less than $\hbar\mathscr{H}_0$. This will allow us to employ the perturbation method of Chapter 3 to calculate relaxation times. However, we see from sections 4.2 and 8.1 that the quadrupole relaxation effect is relatively stronger than those arising from dipolar interactions and anisotropic chemical shift.

9.2 Relaxation times for quadrupole interactions

We compare (9.20) with (3.38). For simplicity we shall omit the superscript 2, since we shall deal only with tensors of rank 2 in this chapter. To keep the notation close to that of (3.38) we write (9.20) as

$$G(t) = \sum_{q=-2}^{2} (-)^q H_{-q}(t) A^{(q)}, \tag{9.21}$$

where

$$H_q(t) = F_q(t), \qquad A^{(q)} = h^{-1} Q_q. \tag{9.22}$$

Then, from (9.18), $A^{(q)}$ is an operator with components

$$A^{(0)} = \alpha[3I_z^2 - I(I+1)]$$

$$A^{(\pm 1)} = \mp \frac{6^{1/2}}{2} \alpha (I_z I_\pm + I_\pm I_z) \tag{9.23}$$

$$A^{(\pm 2)} = \frac{6^{1/2}}{2} \alpha I_\pm^2,$$

where

$$\alpha = \frac{eQ}{2I(2I-1)\hbar}. \tag{9.24}$$

Our first task is to obtain expressions for the time derivatives of $\langle I_z \rangle$ and $\langle I_x \rangle$. This is done by substituting I_z or I_x for I_r in (3.72), taking the values of $A^{(q)}$ from (9.23), summing q from -2 to $+2$ and noting that, as demonstrated in section 4.2, $q = 0, \pm 1, \pm 2$ are associated, respectively, with $j(0), j(\omega_0), j(2\omega_0)$. We find without difficulty that

$$\frac{d\langle I_r \rangle}{dt} = j(0)\langle [A^{(0)}, [I_r, A^{(0)}]] \rangle + j(\omega_0)\langle [A^{(-1)}, [A^{(1)}, I_r]] + [A^{(1)}, [A^{(-1)}, I_r]] \rangle$$

$$+ j(2\omega_0)\langle [A^{(-2)}, [A^{(2)}, I_r]] + [A^{(2)}, [A^{(-2)}, I_r]] \rangle. \tag{9.25}$$

In the course of further calculations we employ the relation

$$[AB, CD] = AC[B, D] + [A, C]DB + A[B, C]D + C[A, D]B \tag{9.26}$$

in place of (4.37) because now B and D are not independent of A and C.

Let us take $I_r \equiv I_z$ in (9.25). The coefficient of $j(0)$ obviously vanishes. By making use of (4.33), (4.36) and (9.26) we readily deduce that

$$[A^{(-1)}, [A^{(1)}, I_z]] = [A^{(1)}, [A^{(-1)}, I_z]]$$

$$= 3\alpha^2 \{-8I_z^3 + 4I(I+1)I_z - I_z\}$$

$$[A^{(-2)}, [A^{(2)}, I_z]] = [A^{(2)}, [A^{(-2)}, I_z]]$$

$$= -6\alpha^2 \{-4I_z^3 + 4I(I+1)I_z - 2I_z\},$$

and hence that

$$\frac{d\langle I_z \rangle}{dt} = -3\alpha^2 \{j(\omega_0)\langle 16I_z^3 - 8I(I+1)I_z + 2I_z \rangle$$

$$+ j(2\omega_0)\langle -16I_z^3 + 16I(I+1)I_z - 8I_z \rangle\}. \tag{9.27}$$

Since in general the right hand side is not proportional to $\langle I_z \rangle$, (9.27) will not lead to an equation of the form (3.67) when the term $\langle I_z \rangle_0 / T_1$ is added as explained in section 4.2.

There are, however, two situations in which an equation like (3.67) is obtained. In the extreme narrowing approximation where $j(\omega_0)$ and $j(2\omega_0)$ in (9.27) are replaced by $j(0)$ we have relaxation with the longitudinal relaxation time T_1 given by

$$\frac{1}{T_1} = 3\alpha^2 [8I(I+1) - 6] j(0), \tag{9.28}$$

that is

$$\frac{1}{T_1} = \frac{3}{2} \left(\frac{eQ}{\hbar} \right)^2 \frac{2I+3}{I^2(2I-1)} j(0) \tag{9.29}$$

by (9.24). We recall that it was proved in the previous section that I must be at least equal to unity for quadrupole interaction to exist. When $I = 1$, we choose the representation (Rose, 1957, pp. 28, 29)

$$I_x = \frac{1}{2^{1/2}} \begin{bmatrix} 0 & 1 & 0 \\ 1 & 0 & 1 \\ 0 & 1 & 0 \end{bmatrix}, \quad I_y = \frac{i}{2^{1/2}} \begin{bmatrix} 0 & -1 & 0 \\ 1 & 0 & -1 \\ 0 & 1 & 0 \end{bmatrix}, \quad I_z = \begin{bmatrix} 1 & 0 & 0 \\ 0 & 0 & 0 \\ 0 & 0 & -1 \end{bmatrix}.$$

$$\tag{9.30}$$

From these we obtain $I_z^3 = I_z$, and it is shown in Appendix A that such an algebraic relation is independent of the chosen representation. When we replace I_z^3 by I_z in (9.27) and put $I = 1$ in (9.24), we deduce (3.67) with

$$\frac{1}{T_1} = \frac{3}{2} \left(\frac{eQ}{\hbar} \right)^2 \{ j(\omega_0) + 4j(2\omega_0) \}. \tag{9.31}$$

If we take $I_r \equiv I_x$ in (9.25), we find that

$$\frac{d\langle I_x \rangle}{dt} = -3\alpha^2 \{ 3j(0) \langle 4I_z^2 I_x + I_x - 4iI_z I_y \rangle$$

$$+ j(\omega_0) \langle I_z^2 I_x + 2I_z I_x I_z + I_x I_z^2 + 2I_+ I_x I_- + 2I_- I_x I_+ \rangle$$

$$+ 2j(2\omega_0) \langle 2I(I+1)I_x - 3I_z^2 I_x - 3I_x I_z^2 \rangle \}, \tag{9.32}$$

which is not of the same form as the first equation of (3.68). In the extreme narrowing case (9.32) becomes

$$\frac{d\langle I_x \rangle}{dt} = -3\alpha^2 \langle 7I_z^2 I_x + [3 + 4I(I+1)]I_x - 12iI_z I_y$$

$$+ 2I_z I_x I_z - 5I_x I_z^2 + 2I_+ I_x I_- + 2I_- I_x I_+ \rangle j(0).$$

On employing (4.36) we deduce that

$$\frac{d\langle I_x \rangle}{dt} = -3\alpha^2 [8I(I+1)-6]j(0)\langle I_x \rangle \tag{9.33}$$

and therefore that the first equation of (3.68) exists. Moreover on comparing (9.33) with (9.28) we see from (9.29) that

$$\frac{1}{T_1} = \frac{1}{T_2} = \frac{3}{2}\left(\frac{eQ}{\hbar}\right)^2 \frac{2I+3}{I^2(2I-1)} j(0). \tag{9.34}$$

When $I = 1$, we deduce from (9.30) that

$$I_+ = 2^{1/2}\begin{bmatrix} 0 & 1 & 0 \\ 0 & 0 & 1 \\ 0 & 0 & 0 \end{bmatrix}, \qquad I_- = 2^{1/2}\begin{bmatrix} 0 & 0 & 0 \\ 1 & 0 & 0 \\ 0 & 1 & 0 \end{bmatrix}, \qquad I_z^2 = \begin{bmatrix} 1 & 0 & 0 \\ 0 & 0 & 0 \\ 0 & 0 & 1 \end{bmatrix}.$$

It follows that

$$I_z^2 I_x = iI_z I_y$$

$$I_z^2 I_x + I_x I_z^2 = I_x, \qquad I_z I_x I_z = 0$$

$$I_+ I_x I_- + I_- I_x I_+ = 2I_x,$$

which when substituted into (9.32) yields

$$\frac{d\langle I_x \rangle}{dt} = -3\alpha^2 \{3j(0) + 5j(\omega_0) + 2j(2\omega_0)\}\langle I_x \rangle. \tag{9.35}$$

We therefore have transverse relaxation with the rate

$$\frac{1}{T_2} = \frac{3}{4}\left(\frac{eQ}{\hbar}\right)^2 \{3j(0) + 5j(\omega_0) + 2j(2\omega_0)\}. \tag{9.36}$$

We conclude that relaxation times T_1 and T_2 do not in general exist for quadrupole relaxation in liquids. They will exist in the extreme narrowing approximation and then T_1 and T_2 will be equal, their common value being given in (9.34). Relaxation times will also exist outside the extreme narrowing approximation, provided that the spin of the nucleus taking part in the relaxation process is equal to unity. Then T_1 and T_2 are given by (9.31) and (9.36), respectively.

9.3 Quadrupole relaxation for molecular models

To complete the calculation of relaxation times, when they exist, we have to evaluate $j(\omega)$. The discussion follows much the same pattern as that of the previous chapter. The $F_q^{(2)}$ of (9.19) are commuting functions of orientational variables only. From (3.48) and (9.22)

$$j(\omega) = \frac{1}{2} \int_{-\infty}^{\infty} \langle F_0(0) F_0(t) \rangle \, e^{-i\omega t} \, dt, \qquad (9.37)$$

where $F_0(t)$ is defined in the laboratory frame S with origin at the centre of mass of the nucleus, in whose quadrupolar relaxation we are interested.

We now transform to the body frame S' with origin at the centre of mass of the molecule containing this nucleus and coordinate axes in the directions of the principal axes of inertia of the molecule. We recall from section 6.1 that we may for our purposes regard such a transformation as a rotation about a common origin. Since V is a scalar, we deduce from (9.19) that the spherical tensor components in S' are given by

$$F_0^{(2)\prime} = \tfrac{1}{2} V_{z'z'}, \qquad F_{\pm 1}^{(2)\prime} = \mp \frac{1}{6^{1/2}} (V_{x'z'} \pm i V_{y'z'})$$

$$F_{\pm 2}^{(2)\prime} = \frac{1}{2(6^{1/2})} (V_{x'x'} - V_{y'y'} \pm 2i V_{x'y'}).$$

The second derivatives constitute the *electric field gradient tensor*

$$\begin{bmatrix} V_{x'x'} & V_{x'y'} & V_{x'z'} \\ V_{y'x'} & V_{y'y'} & V_{y'z'} \\ V_{z'x'} & V_{z'y'} & V_{z'z'} \end{bmatrix},$$

which is real and symmetric. We now take a body coordinate system S'' with origin again at the centre of mass of the above nucleus and in directions that make the field gradient tensor diagonal, so that the only non-vanishing elements are $V_{x''x''}$, $V_{y''y''}$, $V_{z''z''}$. We denote by α, β, γ the Euler angles associated with the transformation from S' to S''. We see from (9.19) that the spherical tensor components in S'' are given by

$$F_0^{(2)\prime\prime} = \tfrac{1}{2} V_{z''z''}, \qquad F_{\pm 1}^{(2)\prime\prime} = 0, \qquad F_{\pm 2}^{(2)\prime\prime} = \frac{1}{2(6^{1/2})} (V_{x''x''} - V_{y''y''}).$$

The x''-, y''-, z''-axes of S'' are chosen such that $|V_{z''z''}| \geqslant |V_{x''x''}| \geqslant |V_{y''y''}|$ and we put

$$\frac{V_{x''x''} - V_{y''y''}}{V_{z''z''}} = \eta, \qquad (9.38)$$

the *asymmetry parameter* for quadrupole interaction. Since all our tensors in this chapter will be of rank two, we omit the superscript 2 and using (9.38) express the tensor components by

$$F_0'' = \tfrac{1}{2} V_{z''z''}, \qquad F_{\pm 1}'' = 0, \qquad F_{\pm 2}'' = \frac{\eta V_{z''z''}}{2(6^{1/2})}. \qquad (9.39)$$

It is customary to write $V_{z''z''}$ as eq. Since $V_{x''z''} = V_{y''z''} = 0$, it follows that eq is the gradient taken in the z''-direction of the electric field at the nucleus; it is in fact the field gradient of greatest magnitude. On comparing (9.39) with (8.21) we find that we can deduce the values of $j(\omega)$ from those in section 8.3 by the substitutions

$$\gamma \delta_{z''} \mapsto eq, \qquad \zeta \mapsto \eta. \tag{9.40}$$

For the spherical model we have from (8.35), (9.40) and (D.24)

$$j(\omega) = \tfrac{1}{20}(1 + \tfrac{1}{3}\eta^2)e^2q^2 \left\{ \frac{G'}{B(G'^2 + \omega'^2)} \right.$$

$$\left. + \frac{6\kappa}{B} \frac{G'(1+G') - \omega'^2}{(G'^2 + \omega'^2)[(1+G')^2 + \omega'^2]} + \cdots \right\} \tag{9.41}$$

$$j(0) = \tfrac{1}{120}(1 + \tfrac{1}{3}\eta^2)e^2q^2 \frac{I_1 B}{kT}(1 + \tfrac{11}{2}\kappa - \tfrac{83}{6}\kappa^2 + \cdots). \tag{9.42}$$

We deduce for spin one from (9.31), (9.36) and (9.41) that

$$\frac{1}{T_1} = \tfrac{3}{40}(1 + \tfrac{1}{3}\eta^2)\left(\frac{e^2qQ}{\hbar}\right)^2 \left\{ \frac{G'}{B(G'^2 + \omega_0'^2)} + \frac{6\kappa}{B} \frac{G'(1+G') - \omega_0'^2}{(G'^2 + \omega_0'^2)[(1+G')^2 + \omega_0'^2]} \right.$$

$$\left. + \frac{4G'}{B(G'^2 + 4\omega_0'^2)} + \frac{24\kappa}{B} \frac{G'(1+G') - 4\omega_0'^2}{(G'^2 + 4\omega_0'^2)[(1+G')^2 + 4\omega_0'^2]} + \cdots \right\} \tag{9.43}$$

$$\frac{1}{T_2} = \tfrac{3}{80}(1 + \tfrac{1}{3}\eta^2)\left(\frac{e^2qQ}{\hbar}\right)^2 \left\{ \frac{3}{BG'} + \frac{18\kappa}{BG'(1+G')} + \frac{5G'}{B(G'^2 + \omega_0'^2)} \right.$$

$$+ \frac{30\kappa}{B} \frac{G'(1+G') - \omega_0'^2}{(G'^2 + \omega_0'^2)[(1+G')^2 + \omega_0'^2]} + \frac{2G'}{B(G'^2 + 4\omega_0'^2)}$$

$$\left. + \frac{12\kappa}{B} \frac{G'(1+G') - 4\omega_0'^2}{(G'^2 + 4\omega_0'^2)[(1+G')^2 + 4\omega_0'^2]} + \cdots \right\}. \tag{9.44}$$

The quantity $e^2qQ/\hbar$ is called the *quadrupole coupling constant*. In the extreme narrowing approximation we find from (9.34) and (9.42)

$$\frac{1}{T_1} = \frac{1}{T_2} = \tfrac{1}{80}(1 + \tfrac{1}{3}\eta^2)\left(\frac{e^2qQ}{\hbar}\right)^2 \frac{2I+3}{I^2(2I-1)} \frac{I_1 B}{kT}(1 + \tfrac{11}{2}\kappa - \tfrac{83}{6}\kappa^2 + \cdots). \tag{9.45}$$

The orientational correlation time is given by (6.5), and so (9.45) may be expressed as

$$\frac{1}{T_1} = \frac{1}{T_2} = \tfrac{3}{40}(1 + \tfrac{1}{3}\eta^2)\left(\frac{e^2qQ}{\hbar}\right)^2 \frac{2I+3}{I^2(2I-1)}\tau_2. \tag{9.46}$$

In the rotational diffusion limit we deduce from (8.38) and (9.40) for the spherical molecule

$$j(\omega) = \tfrac{1}{120}(1 + \tfrac{1}{3}\eta^2) \frac{e^2 q^2 I_1 B}{kT} \frac{1}{1 + \left(\dfrac{I_1 B\omega}{6kT}\right)^2}. \tag{9.47}$$

Then for spin one (9.31), (9.36) and (9.47) yield

$$\frac{1}{T_1} = \tfrac{1}{80}(1 + \tfrac{1}{3}\eta^2)\left(\frac{e^2 qQ}{\hbar}\right)^2 \frac{I_1 B}{kT} \left\{ \frac{1}{1 + \left(\dfrac{I_1 B\omega_0}{6kT}\right)^2} + \frac{4}{1 + \left(\dfrac{I_1 B\omega_0}{3kT}\right)^2} \right\} \tag{9.48}$$

$$\frac{1}{T_2} = \tfrac{1}{160}(1 + \tfrac{1}{3}\eta^2)\left(\frac{e^2 qQ}{\hbar}\right)^2 \frac{I_1 B}{kT} \left\{ 3 + \frac{5}{1 + \left(\dfrac{I_1 B\omega_0}{6kT}\right)^2} + \frac{2}{1 + \left(\dfrac{I_1 B\omega_0}{3kT}\right)^2} \right\}. \tag{9.49}$$

In the extreme narrowing approximation we again obtain (9.46) with τ_2 now given by (6.50).

The asymmetry parameter η is usually quite small; for example, it was found to be 0.007 for benzene (Bonera & Rigamonti, 1965b). Putting $1 + \eta^2/3$ equal to $(1 + \tfrac{1}{6}\eta^2)^2$ we express (9.46) as

$$\frac{1}{T_1} = \frac{1}{T_2} = \frac{3}{40}\left[(1 + \tfrac{1}{6}\eta^2)\frac{e^2 qQ}{\hbar}\right]^2 \frac{2I+3}{I^2(2I-1)}\tau_2.$$

If we take η equal to $\tfrac{1}{3}$, the neglect of $\eta^2/6$ would alter the quantity in the square bracket by a factor 0.02. We shall often neglect η in all the molecular models and approximate (9.46) by

$$\frac{1}{T_1} = \frac{1}{T_2} = \frac{3}{40}\left(\frac{e^2 qQ}{\hbar}\right)^2 \frac{2I+3}{I^2(2I-1)}\tau_2. \tag{9.50}$$

It follows from (D.18) that this result is valid for all molecular models. We see from (9.38) that the vanishing of η makes the z''-axis an axis of cylindrical symmetry for the field gradient tensor. The orientation of the z''-axis with respect to the third coordinate axis of S' is specified by the polar angle β and the azimuthal angle α. Hence we may deduce from (8.44) and (9.40) the result for an asymmetric molecule

$$j(\omega) = \frac{e^2 q^2}{160} \{3 \sin^4 \beta(A + A^*) + 3 \sin^2 2\beta(B + B^*)$$

$$+ (3\cos^2\beta - 1)^2(C + C^*) + 2(6^{1/2})(3\cos^2\beta - 1)\sin^2\beta\cos 2\alpha(D + D^*)$$

$$- 3\sin^2 2\beta\cos 2\alpha(E + E^*) + 3\sin^4\beta\cos 4\alpha(F + F^*)\}. \tag{9.51}$$

The relaxation rates may be obtained from (9.31), (9.34), (9.36) and (9.51).

When the molecule is a symmetric rotator, the values of $j(\omega)$ may be obtained on replacing $\gamma^2 \delta_{z''}^2$ by $e^2 q^2$ in (8.45) for the inertial theory and in

(8.46) for the rotational diffusion theory. In the latter case the extreme narrowing approximation yields

$$j(0) = \frac{e^2q^2}{80} \left\{ \frac{(3\cos^2\beta - 1)^2}{6D_1} + \frac{3\sin^2 2\beta}{5D_1 + D_3} + \frac{3\sin^4\beta}{2D_1 + 4D_3} \right\}.$$

Then, from (9.34),

$$\frac{1}{T_1} = \frac{1}{T_2} = \frac{3}{160} \left(\frac{e^2qQ}{\hbar} \right)^2 \frac{2I+3}{I^2(2I-1)}$$

$$\times \left\{ \frac{(3\cos^2\beta - 1)^2}{6D_1} + \frac{3\sin^2 2\beta}{5D_1 + D_3} + \frac{3\sin^4\beta}{2D_1 + 4D_3} \right\}. \quad (9.52)$$

This is one-half of the common value of the relaxation rates given by Shimizu (1964). Our result agrees with that of Huntress (1968), as we can see by referring back to (6.51) and (6.52). We find from (9.50) and (9.52) that

$$\tau_2 = \frac{(3\cos^2\beta - 1)^2}{24D_1} + \frac{3\sin^2 2\beta}{4(5D_1 + D_3)} + \frac{3\sin^4\beta}{4(2D_1 + 4D_3)}$$

in agreement with (6.48).

Equations (8.48) and (9.40) show that the spectral density for the linear molecule

$$j(\omega) = \frac{e^2q^2}{160} \{ (3\cos^2\beta - 1)^2 \sigma(\omega)_{00} + 3\sin^2 2\beta \sigma(\omega)_{11} + 3\sin^4\beta \sigma(\omega)_{22} \}. \quad (9.53)$$

The values of T_1^{-1}, T_2^{-1} obtained for spin one from (9.31), (9.36), (9.53) and (D.40)–(D.42) are lengthy. If we suppose that the angle β between the z''- and z'-axes vanishes, so that the two axes coincide, (9.53) becomes

$$j(\omega) = \frac{e^2q^2\sigma(\omega)_{00}}{40} \quad (9.54)$$

$$= \frac{e^2q^2}{20B} \left\{ \frac{(1 + 6\kappa + 33\kappa^2 + \cdots)G_0}{G_0^2 + \omega'^2} - \frac{(6\kappa + 48\kappa^2 + \cdots)(1 + G_0)}{(1 + G_0)^2 + \omega'^2} + \cdots \right\}.$$

In the rotational diffusion limit $G_0 = 6\kappa$ by (D.34), so that $BG_0 = 6kT/(I_1 B)$ and (9.54) reduces to

$$j(\omega) = \frac{e^2q^2}{120} \frac{I_1 B}{kT} \frac{1}{1 + \left(\dfrac{I_1 B\omega}{6kT} \right)^2},$$

which is the value for $\eta = 0$ of $j(\omega)$ given by (9.47) for the spherical molecule.

We conclude that, when $\beta = 0$, the relaxation times for linear molecules with spin one in rotational diffusion theory are given by (9.48) and (9.49) with $\eta = 0$. As shown in section 8.3 for anisotropic chemical shift, the same result is true in the case of the symmetric rotator molecule when the z''-axis is parallel to the axis of rotational symmetry. Hence, in all three cases the relaxation times for spin one nuclei in rotational diffusion theory are given by

$$\frac{1}{T_1} = \frac{1}{80}\left(\frac{e^2 qQ}{\hbar}\right)^2 \frac{I_1 B}{kT} \left\{ \frac{1}{1+\left(\dfrac{I_1 B\omega_0}{6kT}\right)^2} + \frac{4}{1+\left(\dfrac{I_1 B\omega_0}{3kT}\right)^2} \right\} \qquad (9.55)$$

$$\frac{1}{T_2} = \frac{1}{160}\left(\frac{e^2 qQ}{\hbar}\right)^2 \frac{I_1 B}{kT} \left\{ 3 + \frac{5}{1+\left(\dfrac{I_1 B\omega_0}{6kT}\right)^2} + \frac{2}{1+\left(\dfrac{I_1 B\omega_0}{3kT}\right)^2} \right\} \cdot \quad (9.56)$$

In the extreme narrowing approximation the relaxation times for any spin are given by (9.50), where τ_2 has the value $I_1 B/(6kT)$.

10

Relaxation by spin-rotational interaction

10.1 The spin-rotational Hamiltonian

When a molecule in a liquid rotates, the motion of the electrons in the molecule generates a magnetic field, which is experienced by the molecular nuclei. Thus, for example, a flux density of about 10^{-3} T is experienced by the hydrogen nucleus in the HCl molecule (Farrar & Becker, 1971, p. 63). Hence, the molecular rotation produces a time dependent field that is a small perturbation of a laboratory field, which produces a flux density 0.1–1 T. The effect is clearly intramolecular. If $\mathbf{I}$ is the spin operator of a nucleus and $\hbar\mathbf{J}$ the angular momentum operator of the molecule, we may express the Hamiltonian $\hbar G(t)$ for the interaction between the spin and the perturbing field by

$$\hbar G(t) = \hbar\mathbf{I} \cdot \mathbf{C} \cdot \mathbf{J}, \qquad (10.1)$$

where $\mathbf{C}$ is a three-by-three real tensor called the *spin-rotation tensor*. We shall consider the contribution to nuclear magnetic relaxation of identical nuclei in identical molecules, and for this purpose it will be adequate to study the interaction expressed by (10.1). The importance of spin-rotational interactions as a relaxation mechanism in liquids was first recognized by Gutowsky, Lawrenson & Shimomura (1961). A relaxation theory for spin $\frac{1}{2}$ nuclei in spherical molecules was first proposed by Hubbard (1963b).

It would not be easy to set up a theory that would provide specific values of the components of the spin-rotation tensor. We shall therefore work from (10.1) and examine its implications for relaxation, which we shall later compare with experimental findings. Let us write

$$G(t) = \sum_{\mu,\nu=1}^{3} I'_{\mu}C'_{\mu\nu}J'_{\nu}, \qquad (10.2)$$

where the primes indicate that the components of $\mathbf{I}$, $\mathbf{C}$ and $\mathbf{J}$ are referred to a molecular frame of coordinate axes; to be more precise, we take the origin at the molecular centre of mass and the third coordinate axis passing through the nucleus under investigation. Leaving J'_{ν} unaltered we replace the summation over μ by a summation over spherical tensor components labelled $-1, 0, 1$, so that

$$\sum_{\mu=1}^{3} I'_\mu C'_{\mu\nu} = \sum_{m=-1}^{1} I'_m b'_{m\nu}; \tag{10.3}$$

that is,

$$I'_3 C'_{3\nu} + I'_1 C'_{1\nu} + I'_2 C'_{2\nu} = I'_0 b'_{0\nu} + I'_{+1} b'_{1\nu} + I'_{-1} b'_{-1\nu}. \tag{10.4}$$

We have written I'_{+1} on the right hand side for the spherical component in order to distinguish it from the cartesian component I'_1 on the left hand side. From (B.29)

$$I'_{+1} = -\frac{I'_1 + iI'_2}{2^{1/2}}, \qquad I'_{-1} = \frac{I'_1 - iI'_2}{2^{1/2}}, \qquad I'_0 = I'_3, \tag{10.5}$$

so that

$$I'_1 = -\frac{I'_{+1} - I'_{-1}}{2^{1/2}}, \qquad I'_2 = \frac{i(I'_{+1} + I'_{-1})}{2^{1/2}}. \tag{10.6}$$

On substituting for I'_1, I'_2, I'_3 from (10.5) and (10.6) into the left hand side of (10.4) and comparing coefficients of I'_0, I'_{+1}, I'_{-1} we deduce that

$$b'_{0\nu} = C'_{3\nu}, \qquad b'_{\pm 1,\nu} = \mp \frac{C'_{1\nu} \mp iC'_{2\nu}}{2^{1/2}}. \tag{10.7}$$

We note that

$$b'_{-m\nu} = (-)^m b'^*_{m\nu} \tag{10.8}$$

but that nevertheless, as we see on comparing (10.7) with (10.5), $b'_{-1\nu}$, $b'_{0\nu}$, $b'_{1\nu}$ are not spherical tensor components.

In our treatment of nuclear magnetic relaxation problems we have expressed the part of $G(t)$ that is a function of nuclear spins in terms of variables related to a laboratory coordinate system, which we write without primes. In the present case the laboratory system is that with its origin at the molecular centre of mass and with the directions of the coordinate axes fixed with respect to the laboratory. If at time t the orientation of the molecule with reference to the laboratory system is specified by the Euler angles $\alpha(t)$, $\beta(t)$, $\gamma(t)$, (B.17) gives

$$I'_m = \sum_{k=-1}^{1} D^1_{km}(\alpha(t), \beta(t), \gamma(t)) I_k. \tag{10.9}$$

On combining (10.2), (10.3) and (10.9) we have

$$G(t) = \sum_{k=-1}^{1} I_k \sum_{\nu=1}^{3} \sum_{m=-1}^{1} b'_{m\nu} D^1_{km}(\alpha(t), \beta(t), \gamma(t)) J'_\nu. \tag{10.10}$$

10.2 The spin-rotational relaxation times

We relate the results of the previous section to those of section 3.3. Equation (10.10) is expressible as (3.38) with $j = 1$, namely,

$$G(t) = \sum_{q=-1}^{1} (-)^q H_{-q}(t) A^{(q)} \tag{10.11}$$

with

$$A^{(q)} = I_q \tag{10.12}$$

$$(-)^q H_{-q}(t) = \sum_{v=1}^{3} \sum_{m=-1}^{1} b'_{mv} D^1_{qm}(\alpha(t), \beta(t), \gamma(t)) J'_v. \tag{10.13}$$

Since for physical reasons the interaction Hamiltonian of (10.1) is invariant for rotations of the laboratory frame of reference and since I_q is the component of a spherical tensor of rank one, it follows from Appendix B and (10.11) that the right hand side of (10.13) is the adjoint of the qth component of a spherical tensor of rank one. Since J'_v is self-adjoint, the complex conjugate of the qth component is equal to its adjoint. Hence $H_q(t)$ defined by

$$H_q(t) = \sum_{v=1}^{3} \sum_{m=-1}^{1} b'^*_{mv} D^{1*}_{qm}(\alpha(t), \beta(t), \gamma(t)) J'_v \tag{10.14}$$

is the qth component of a spherical tensor of rank one.

At this stage we make two assumptions:

(a) the third axis of the molecular frame, which passes through the nucleus, is a principal axis of inertia of the molecule;

(b) the motion of the molecule may be treated classically.

If the first assumption is not valid, it will be necessary to work with two different molecular sets of axes, as explained in section 8.1 for the discussion of relaxation by anisotropic chemical shift. Accepting the first assumption we take the other two principal axes of inertia through the centre of mass as the first and second coordinate axes. Then the second assumption allows us to replace $\hbar J'_v$ at time t by $I_v \omega_v(t)$, where I_1, I_2, I_3 are the principal moments of inertia and $\omega_1(t), \omega_2(t), \omega_3(t)$ the corresponding components of angular velocity. With these replacements (10.14) becomes

$$H_q(t) = \hbar^{-1} \sum_{v=1}^{3} \sum_{m=-1}^{1} b'^*_{mv} I_v D^{1*}_{qm}(\alpha(t), \beta(t), \gamma(t)) \omega_v(t). \tag{10.15}$$

We see that $H_q(t)$ is now independent of quantum mechanical operators and so is a commuting quantity. However, $H_q(t)$ is a function of both orientational and angular velocity variables, so (C.32) cannot be applied to it.

The dependence of $H_q(t)$ on angular velocity variables does not invalidate results based on (3.41) and the steady state of the motion. Thus

$$\langle H_q^*(0)H_q(t)\rangle = \delta_{qq'}\langle H_0(0)H_0(t)\rangle \qquad (10.16)$$

and $C(t)$ defined by

$$C(t)=\langle H_0(0)H_0(t)\rangle \qquad (10.17)$$

is a real and even function of t. For calculational purposes it will be convenient to employ $c(s)$, the Laplace transform of $C(t)$, which from (10.17) is given by

$$c(s)= \int_0^\infty e^{-st}\langle H_0(0)H_0(t)\rangle \, dt. \qquad (10.18)$$

It follows from (3.48) that the spectral density

$$j(\omega)=\tfrac{1}{2}[c(i\omega)+c(-i\omega)]. \qquad (10.19)$$

Let us express the longitudinal and transverse relaxation times T_1 and T_2 in terms of the spectral density $j(\omega)$. From (10.12) and (B.29)

$$A^{(-1)}=\frac{I_-}{2^{1/2}}, \qquad A^{(0)}=I_z, \qquad A^{(1)}=-\frac{I_+}{2^{1/2}}, \qquad (10.20)$$

where

$$I_+=I_x+iI_y, \qquad I_-=I_x-iI_y.$$

We employ (3.72) together with the results established in section 4.2 that $j(0)$ arises from $q=0$ and that $j(\omega_0)$ arises from $q=\pm 1$. The coefficient of $j(0)$ in $d\langle I_r\rangle/dt$,

$$\langle 2I_zI_rI_z-I_z^2I_r-I_rI_z^2\rangle=\langle I_z[I_r,I_z]+[I_z,I_r]I_z\rangle,$$

which vanishes for $I_r=I_z$ and equals $-\langle I_x\rangle$ for $I_r=I_x$. Similarly we find that the coefficient of $j(\omega_0)$ in $d\langle I_r\rangle/dt$ is

$$\tfrac{1}{2}\langle [[I_+,I_r],I_-]+[[I_-,I_r],I_+]\rangle,$$

which for $I_r=I_z$ equals $-2\langle I_z\rangle$ and for $I_r=I_x$ equals $-\langle I_x\rangle$, by (4.36). It follows that

$$\frac{1}{T_1}=2j(\omega_0), \qquad \frac{1}{T_2}=j(0)+j(\omega_0), \qquad (10.21)$$

and in the extreme narrowing approximation

$$\frac{1}{T_1}=\frac{1}{T_2}=2j(0). \qquad (10.22)$$

10.3 Calculation of spectral densities

We deduce from (B.3), (B.23) and (10.15) that

$$H_0(t) = h^{-1} \sum_{v=1}^{3} \sum_{m=-1}^{1} b_{mv}^{\prime *} D_{0m}^{1*}(\alpha(t), \beta(t), \gamma(t)) I_v \omega_v(t)$$

$$= \left(\frac{4\pi}{3h^2}\right)^{1/2} \sum_{v=1}^{3} \sum_{m=-1}^{1} (-)^m b_{mv}^{\prime *} Y_{1,-m}^*(\beta(t), \gamma(t)) I_v \omega_v(t),$$

so that

$$H_0(0)H_0(t) = \frac{4\pi}{3h^2} \sum_{\mu,v=1}^{3} \sum_{m,n=-1}^{1} (-)^m b_{mv}^{\prime *} b_{n\mu}^{\prime *} I_\mu I_v Y_{1,-m}^*(\beta(0), \gamma(0))$$

$$\times Y_{1n}(\beta(t), \gamma(t)) \omega_\mu(t) \omega_v(0).$$

We then deduce from (10.17) that

$$C(t) = \frac{4\pi}{3h^2} \sum_{\mu,v=1}^{3} \sum_{m,n=-1}^{1} (-)^m b_{mv}^{\prime *} b_{n\mu}^{\prime *} I_\mu I_v$$

$$\times \langle Y_{1,-m}^*(\beta(0), \gamma(0)) Y_{1n}(\beta(t), \gamma(t)) \omega_\mu(t) \omega_v(0) \rangle. \qquad (10.23)$$

We have an ensemble average of a product of orientational and angular velocity variables. Since the rotation operator which determines the orientation satisfies the differential equation (B.12) and since this equation contains angular velocity variables, the orientational and angular velocity variables are not mutually independent. Thus one cannot equate the ensemble average of the above product to the product of the ensemble averages over the orientational variables and over the angular velocity variables.

We define the rotation operator $R(t)$ as that which brings the molecular coordinate system from its orientation at time zero to its orientation at time t. Then (B.16) gives

$$Y_{1s}(\beta(0), \gamma(0)) = \sum_{k=-1}^{1} R(t)_{ks}^1 Y_{1k}(\beta(t), \gamma(t)),$$

and conversely from (B.15)

$$Y_{1s}(\beta(t), \gamma(t)) = \sum_{k=-1}^{1} R^+(t)_{ks}^1 Y_{1k}(\beta(0), \gamma(0)).$$

It follows from (10.23) that

$$C(t) = \frac{4\pi}{3h^2} \sum_{\mu,v=1}^{3} I_\mu I_v \sum_{m,n=-1}^{1} (-)^m b_{mv}^{\prime *} b_{n\mu}^{\prime *}$$

$$\times \left\langle Y_{1,-m}^*(\beta(0), \gamma(0)) \sum_{k=-1}^{1} R^+(t)_{kn}^1 Y_{1k}(\beta(0), \gamma(0)) \omega_\mu(t) \omega_v(0) \right\rangle. \qquad (10.24)$$

To calculate the ensemble average we follow the procedure introduced in Appendix D of first averaging with the initial condition that $R(0)$ is the identity operator, and then averaging over the initial angle variables $\beta(0), \gamma(0)$ by employing the probability density function $(4\pi)^{-1} \sin \beta(0) \, d\beta(0) \, d\gamma(0)$. Thus

$$\left\langle Y^*_{1,-m}(\beta(0),\gamma(0)) \sum_{k=-1}^{1} R^+(t)_{kn}^{1} Y_{1k}(\beta(0),\gamma(0))\omega_\mu(t)\omega_\nu(0) \right\rangle$$

$$= (4\pi)^{-1} \sum_{k=-1}^{1} \int_0^{2\pi} d\gamma(0) \int_0^{\pi} d\beta(0) \sin \beta(0) Y^*_{1,-m}(\beta(0),\gamma(0))$$

$$\times \langle R^+(t)\omega_\mu(t)\omega_\nu(0)\rangle_{kn} Y_{1k}(\beta(0),\gamma(0))$$

$$= (4\pi)^{-1}\langle R^+(t)\omega_\mu(t)\omega_\nu(0)\rangle_{-m,n} = (4\pi)^{-1}\langle R(t)\omega_\mu(t)\omega_\nu(0)\rangle^*_{n,-m},$$

where the subscripts denote the $n, -m$ element in the three-dimensional representation with basis $Y_{1,-1}(\beta(0),\gamma(0))$, $Y_{10}(\beta(0),\gamma(0))$, $Y_{11}(\beta(0),\gamma(0))$.

We return to (10.24) recalling that $C(t)$, being a real scalar, is equal to $C(t)^*$, and we therefore deduce that

$$C(t)=\frac{1}{3\hbar^2} \sum_{\mu,\nu=1}^{3} I_\mu I_\nu \sum_{m,n=-1}^{1} (-)^m b'_{m\nu} b'_{n\mu}\langle R(t)\omega_\mu(t)\omega_\nu(0)\rangle_{n,-m}. \quad (10.25)$$

It will be understood in this chapter that angle brackets signify an average calculated with the initial condition that $R(0)$ is the identity.

We deduce from (10.18) and (10.25) that

$$c(s)=\frac{1}{3\hbar^2} \sum_{\mu,\nu=1}^{3} \sum_{m,n=-1}^{1} (-)^m b'_{n\mu} b'_{m\nu} I_\mu I_\nu \left(\int_0^\infty e^{-st}\langle R(t)\omega_\mu(t)\omega_\nu(0)\rangle \, dt \right)_{n,-m}.$$

$$\quad (10.26)$$

If we have calculated the integral, we can deduce the value of $c(s)$ and thence the spectral density $j(\omega)$ from (10.19). The relaxation rates may be obtained from (10.21) and (10.22). We see that in the extreme narrowing approximation

$$\frac{1}{T_1}=\frac{1}{T_2}=2c(0). \quad (10.27)$$

The *spin-rotational correlation time* τ_{sr} is defined as the integral from zero to infinity of the normalized autocorrelation function of $H_q(t)$. According to (10.16) and (10.17) the latter is $C(t)/C(0)$ and hence from (10.18)

$$\tau_{sr}=\frac{c(0)}{C(0)}. \quad (10.28)$$

From (10.25)

$$C(0) = \frac{1}{3\hbar^2} \sum_{\mu,\nu=1}^{3} \sum_{m,n=-1}^{1} (-)^m b'_{n\mu} b'_{m\nu} I_\mu I_\nu \delta_{n,-m} \langle \omega_\mu(0)\omega_\nu(0)\rangle.$$

For a body of any shape (McConnell, 1980b, section 11.4)

$$\langle \omega_\mu(0)\omega_\nu(0)\rangle = \frac{kT\delta_{\mu\nu}}{I_\mu},$$

and so

$$C(0) = \frac{kT}{3\hbar^2} \sum_{\mu=1}^{3} \sum_{m=-1}^{1} (-)^m b'_{m\mu} b'_{-m\mu} I_\mu$$

$$\tau_{sr} = \frac{3\hbar^2}{kT} \frac{c(0)}{\sum_{\mu=1}^{3} \sum_{m=-1}^{1} (-)^m b'_{m\mu} b'_{-m\mu} I_\mu}, \tag{10.29}$$

where $c(0)$ is to be obtained from (10.26).

It is seen from (10.26) that the central mathematical problem is to calculate $\langle R(t)\omega_\mu(t)\omega_\nu(0)\rangle$. The method of finding $R(t)$ is outlined in Appendix D. The solutions of the Euler–Langevin equations are centred random variables, but in general they are non-Gaussian (McConnell, 1980b, section 11.1). This has the consequence that the mean value of the product of an uneven number of ω_i's may not vanish. Moreover, the infinitesimal generators J_1, J_2, J_3 related to the rotating molecular coordinate system obey

$$[J_\mu, J_\nu] = -i(\mathbf{J} \cdot \mathbf{e}_\mu \times \mathbf{e}_\nu) \tag{10.30}$$

(Van Vleck, 1951).

The following values of $\varepsilon^n F^{(n)}(t)$, $\varepsilon^n \Omega^{(n)}(t)$ will be required for the calculations of the present chapter (McConnell, 1982a):

$$\varepsilon F^{(1)}(t) = -i \int_0^t (\mathbf{J} \cdot \boldsymbol{\omega}(t_1))\, dt_1$$

$$\varepsilon^2 F^{(2)}(t) = -\int_0^t dt_1 \int_0^{t_1} dt_2 [(\mathbf{J}\cdot\boldsymbol{\omega}(t_1))(\mathbf{J}\cdot\boldsymbol{\omega}(t_2)) - \langle(\mathbf{J}\cdot\boldsymbol{\omega}(t_1))(\mathbf{J}\cdot\boldsymbol{\omega}(t_2))\rangle]$$

$$\begin{aligned}
\varepsilon^3 F^{(3)}(t) = i \int_0^t dt_1 \int_0^{t_1} dt_2 \int_0^{t_2} dt_3 & [(\mathbf{J}\cdot\boldsymbol{\omega}(t_1))(\mathbf{J}\cdot\boldsymbol{\omega}(t_2))(\mathbf{J}\cdot\boldsymbol{\omega}(t_3)) \\
& - (\mathbf{J}\cdot\boldsymbol{\omega}(t_1))\langle(\mathbf{J}\cdot\boldsymbol{\omega}(t_2))(\mathbf{J}\cdot\boldsymbol{\omega}(t_3))\rangle \\
& - (\mathbf{J}\cdot\boldsymbol{\omega}(t_2))\langle(\mathbf{J}\cdot\boldsymbol{\omega}(t_1))(\mathbf{J}\cdot\boldsymbol{\omega}(t_3))\rangle \\
& - (\mathbf{J}\cdot\boldsymbol{\omega}(t_3))\langle(\mathbf{J}\cdot\boldsymbol{\omega}(t_1))(\mathbf{J}\cdot\boldsymbol{\omega}(t_2))\rangle \\
& - \langle(\mathbf{J}\cdot\boldsymbol{\omega}(t_1))(\mathbf{J}\cdot\boldsymbol{\omega}(t_2))(\mathbf{J}\cdot\boldsymbol{\omega}(t_3))\rangle]
\end{aligned} \tag{10.31}$$

$$\varepsilon^4 F^{(4)}(t) = -\varepsilon^4 \int_0^t \Omega^{(4)}(t_1)\,dt_1 - i\varepsilon^3 \int_0^t (\mathbf{J} \cdot \boldsymbol{\omega}(t_1)) F^{(3)}(t_1)\,dt_1$$

$$-\varepsilon^4 \int_0^t \{F^{(1)}(t_1)\Omega^{(3)}(t_1) + F^{(2)}(t_1)\Omega^{(2)}(t_1)\}\,dt_1;$$

$$\varepsilon\Omega^{(1)}(t) = 0, \qquad \varepsilon^2\Omega^{(2)}(t) = -\int_0^t \langle(\mathbf{J} \cdot \boldsymbol{\omega}(t))(\mathbf{J} \cdot \boldsymbol{\omega}(t_1))\rangle\,dt_1$$

$$\varepsilon^3\Omega^{(3)}(t) = i\int_0^t dt_1 \int_0^{t_1} dt_2 \langle(\mathbf{J} \cdot \boldsymbol{\omega}(t))(\mathbf{J} \cdot \boldsymbol{\omega}(t_1))(\mathbf{J} \cdot \boldsymbol{\omega}(t_2))\rangle$$

$$\varepsilon^4\Omega^{(4)}(t) = \int_0^t dt_1 \int_0^{t_1} dt_2 \int_0^{t_2} dt_3 \{\langle(\mathbf{J} \cdot \boldsymbol{\omega}(t))(\mathbf{J} \cdot \boldsymbol{\omega}(t_1))(\mathbf{J} \cdot \boldsymbol{\omega}(t_2))(\mathbf{J} \cdot \boldsymbol{\omega}(t_3))\rangle$$

$$-\langle(\mathbf{J} \cdot \boldsymbol{\omega}(t))(\mathbf{J} \cdot \boldsymbol{\omega}(t_1))\rangle\langle(\mathbf{J} \cdot \boldsymbol{\omega}(t_2))(\mathbf{J} \cdot \boldsymbol{\omega}(t_3))\rangle \qquad (10.32)$$

$$-\langle(\mathbf{J} \cdot \boldsymbol{\omega}(t))(\mathbf{J} \cdot \boldsymbol{\omega}(t_2))\rangle\langle(\mathbf{J} \cdot \boldsymbol{\omega}(t_1))(\mathbf{J} \cdot \boldsymbol{\omega}(t_3))\rangle$$

$$-\langle(\mathbf{J} \cdot \boldsymbol{\omega}(t))(\mathbf{J} \cdot \boldsymbol{\omega}(t_3))\rangle\langle(\mathbf{J} \cdot \boldsymbol{\omega}(t_1))(\mathbf{J} \cdot \boldsymbol{\omega}(t_2))\rangle\}.$$

On substituting (10.32) into (D.2) and integrating we obtain $\langle R(t)\rangle$. Then from (D.1) we have

$$\langle R(t)\omega_\mu(t)\omega_\nu(0)\rangle = \langle(E + \varepsilon F^{(1)}(t) + \varepsilon^2 F^{(2)}(t) + \varepsilon^3 F^{(3)}(t)$$

$$+ \varepsilon^4 F^{(4)}(t) + \cdots)\omega_\mu(t)\omega_\nu(0)\rangle\langle R(t)\rangle. \qquad (10.33)$$

The above results are true for a rigid molecule of any shape. To derive analytical expressions for relaxation and correlation times we choose special molecular models.

10.4 Spin-rotational relaxation for molecular models

We apply the theory of the previous section to the spherical model of the molecule. Then the principal moments of inertia are equal and the Euler–Langevin equations (2.57) reduce to

$$I_1\dot{\omega}(t) = -I_1 B\omega(t) + \mathbf{A}(t). \qquad (10.34)$$

The components of angular velocity are centred Gaussian random variables obeying (McConnell, 1980b, section 9.2)

$$\langle\omega_i(t_k)\omega_j(t_l)\rangle = \frac{\delta_{ij}kT}{I_1} e^{-B|t_k - t_l|}. \qquad (10.35)$$

We see from (10.31)–(10.33) that $\varepsilon^3\Omega^{(3)}(t)$ vanishes and that there is no contribution to $\langle R(t)\omega_\mu(t)\omega_\nu(0)\rangle$ from the terms $\varepsilon F^{(1)}(t)$, $\varepsilon^3 F^{(3)}(t)$ in the series because they yield only averages of products of an odd number of

ω_i's. Since the calculations for the spherical model have already been presented in some detail (McConnell, 1982*a*), we shall limit the discussion to the main steps. It is found from (10.32) and (10.35) that

$$\varepsilon^2 \Omega^{(2)}(t) = \frac{kT}{I_1 B} J^2 (1 - e^{-Bt})$$

$$\varepsilon^4 \Omega^{(4)}(t) = -\left(\frac{kT}{I_1}\right)^2 J^2 \int_0^t dt_1 \int_0^{t_1} dt_2 \int_0^{t_2} dt_3 \, e^{-B(t+t_1-t_2-t_3)}.$$

(10.36)

From (10.31) it is deduced that

$$\varepsilon^2 F^{(2)}(t) = -\int_0^t dt_1 \int_0^{t_1} dt_2 \left[\sum_{r,s=1}^{3} J_r J_s \omega_r(t_1) \omega_s(t_2) - \frac{kT}{I_1} J^2 e^{-B(t_1-t_2)} \right]$$

$$\varepsilon^3 F^{(3)}(t) = i \int_0^t dt_1 \int_0^{t_1} dt_2 \int_0^{t_2} dt_3 \left\{ \sum_{b,c,d=1}^{3} J_b J_c J_d \omega_b(t_1) \omega_c(t_2) \omega_d(t_3) \right.$$

$$- \frac{kT}{I_1} J^2 \left[\sum_{b=1}^{3} J_b \omega_b(t_1) e^{-B(t_2-t_3)} + \sum_{c=1}^{3} J_c \omega_c(t_2) e^{-B(t_1-t_3)} \right.$$

$$\left. \left. + \sum_{d=1}^{3} J_d \omega_d(t_3) e^{-B(t_1-t_2)} \right] \right\}.$$

From these, (10.31) and (10.36) a lengthy expression for $\varepsilon^4 F^{(4)}(t)$ may be obtained.

We take the value of $\langle R(t) \rangle$ from (D.20). Substituting the values of $\varepsilon^2 F^{(2)}(t)$, $\varepsilon^4 F^{(4)}(t)$, $\langle R(t) \rangle$ into (10.33) we deduce the value of the operator $\langle R(t) \omega_\mu(t) \omega_\nu(0) \rangle$. Finally we employ the theory of Laplace transforms (McConnell, 1980*b*, Appendix C) to obtain

$$\int_0^\infty e^{-st} \langle R(t) \omega_\mu(t) \omega_\nu(0) \rangle \, dt$$

$$= \delta_{\mu\nu} E \frac{kT}{I_1} \left\{ \frac{E}{s+B+BG_j} + \frac{\kappa B J^2}{(s+B+BG_j)(s+2B+BG_j)} \right.$$

$$+ \kappa^2 \left[\frac{J^4/2 + 5J^2/4}{s+B+BG_j} - \frac{J^4+J^2}{s+2B+BG_j} - \frac{BJ^2}{(s+2B+BG_j)^2} \right.$$

$$+ \frac{J^4/2 - J^2/4}{s+3B+BG_j} + \frac{B^4 J^2}{(s+B+BG_j)^3(s+2B+BG_j)^2}$$

$$\left. \left. + \frac{B^4 J^2}{(s+B+BG_j)^2(s+2B+BG_j)^2(s+3B+BG_j)} \right] \right\}$$

$$+ i(\mathbf{J} \cdot \mathbf{e}_\mu \times \mathbf{e}_\nu) \frac{\kappa kT}{I_1} \left\{ \frac{E}{s+B+BG_j} - \frac{BE}{(s+B+BG_j)^2} - \frac{E}{s+2B+BG_j} \right.$$

$$+\kappa\left[J^2\left(\frac{1/2}{s+B+BG_j}-\frac{B^2}{(s+B+BG_j)^3}-\frac{1}{s+2B+BG_j}+\frac{1/2}{s+3B+BG_j}\right)\right.$$

$$+\frac{B^4(J^2-E)}{(s+B+BG_j)^3(s+2B+BG_j)^2} \tag{10.37}$$

$$\left.\left.+\frac{B^4(2J^2-4E)}{(s+B+BG_j)^2(s+2B+BG_j)^2(s+3B+BG_j)}\right]\right\}$$

$$+J_\mu J_\nu\frac{\kappa kT}{I_1}\left\{-\frac{E+2\kappa J^2+\kappa^2[3J^4/2+5J^2/4]}{s+BG_j}+\frac{2E+3\kappa J^2/2}{s+B+BG_j}\right.$$

$$+\frac{3\kappa BJ^2}{(s+B+BG_j)^2}-\frac{E}{s+2B+BG_j}+\frac{\kappa J^2/2}{s+3B+BG_j}$$

$$+\kappa B^4\left[\frac{2J^2-2E+2\kappa(J^4-J^2)}{(s+BG_j)(s+B+BG_j)^3(s+2B+BG_j)}\right.$$

$$+\frac{2J^2-3E}{(s+B+BG_j)^3(s+2B+BG_j)^2}$$

$$\left.\left.+\frac{2J^2-5E}{(s+B+BG_j)^2(s+2B+BG_j)^2(s+3B+BG_j)}\right]\right\}$$

$$+\cdots,$$

where

$$G_j=j(j+1)\kappa\left\{1+\tfrac{1}{2}\kappa+\tfrac{7}{12}\kappa^2+\left[\frac{17}{18}-\frac{j(j+1)}{8}\right]\kappa^3+\cdots\right\}$$

$$\kappa=\frac{kT}{I_1B^2}. \tag{10.38}$$

In the present problem $j=1$, so that $J^2=2E$. It follows that the quantities within the braces on the right hand side of (10.37) are multiples of the identity.

For the sphere (10.26) simplifies to

$$c(s)=\frac{I_1^2}{3\hbar^2}\sum_{\mu,\nu=1}^{3}\sum_{m,n=-1}^{1}(-)^m b'_{n\mu}b'_{m\nu}\left(\int_0^\infty e^{-st}\langle R(t)\omega_\mu(t)\omega_\nu(0)\rangle\,dt\right)_{n,-m} \tag{10.39}$$

In order to evaluate the sum we write explicit expressions for the matrix representatives of J_1, J_2, J_3 satisfying (10.30) when the basis of the representation is $Y_{1,-1}, Y_{10}, Y_{11}$:

$$J_1=\frac{1}{2^{1/2}}\begin{bmatrix}0&1&0\\1&0&1\\0&1&0\end{bmatrix},\quad J_2=\frac{i}{2^{1/2}}\begin{bmatrix}0&-1&0\\1&0&-1\\0&1&0\end{bmatrix},\quad J_3=\begin{bmatrix}-1&0&0\\0&0&0\\0&0&1\end{bmatrix}.$$

$$\tag{10.40}$$

Let us assume that the third body axis is an axis of symmetry C_n, that is the body is symmetric under a rotation through an angle $2\pi/n$ about the axis, and let us suppose that n is an integer not less than 3. Then we may write (cf. Townes & Schawlow, 1975, pp. 219, 220)

$$C'_{33}=C_\parallel, \qquad C'_{11}=C'_{22}=C_\perp, \qquad C'_{pq}=0 \qquad (p\neq q). \quad (10.41)$$

It follows from (10.7) that

$$b'_{03}=C_\parallel, \qquad b'_{\pm1,1}=\mp\frac{C_\perp}{2^{1/2}}, \qquad b'_{\pm1,2}=\frac{iC_\perp}{2^{1/2}}$$

$$b'_{01}=b'_{02}=b'_{\pm1,3}=0. \quad (10.42)$$

It may now be readily deduced from (10.40) and (10.42) that

$$\sum_{\mu,\nu=1}^{3}\sum_{m,n=-1}^{1}(-)^m b'_{n\mu}b'_{m\nu}\delta_{\mu\nu}E_{n,-m}=C_\parallel^2+2C_\perp^2 \quad (10.43)$$

$$\sum_{\mu,\nu=1}^{3}\sum_{m,n=-1}^{1}(-)^m b'_{n\mu}b'_{m\nu}[i(\mathbf{J}\cdot\mathbf{e}_\mu\times\mathbf{e}_\nu)]_{n,-m}=-2C_\perp^2-4C_\perp C_\parallel \quad (10.44)$$

$$\sum_{\mu,\nu=1}^{3}\sum_{m,n=-1}^{1}(-)^m b'_{n\mu}b'_{m\nu}(J_\mu J_\nu)_{n,-m}=0. \quad (10.45)$$

Equation (10.45) shows that the $J_\mu J_\nu$-term of (10.37) gives no contribution to $c(s)$.

We have now all the information required to derive explicit expressions for T_1, T_2 from (10.19), (10.21), (10.37) and (10.39). Since it would be cumbersome to write out the expressions in full, we shall confine our attention to the extreme narrowing case. Putting $j=1, s=0$ in (10.37) and (10.38), and expanding the results as power series in κ we obtain

$$\lim_{s\to0}\int_0^\infty e^{-st}\langle R(t)\omega_\mu(t)\omega_\nu(0)\rangle_{j=1}\,dt$$

$$=\frac{kT}{I_1 B}\left\{\delta_{\mu\nu}E+\kappa\left[-\delta_{\mu\nu}E-\frac{i}{2}(\mathbf{J}\cdot\mathbf{e}_\mu\times\mathbf{e}_\nu)\right]\right.$$

$$+\kappa^2[\tfrac{13}{6}\delta_{\mu\nu}E+\tfrac{13}{12}i(\mathbf{J}\cdot\mathbf{e}_\mu\times\mathbf{e}_\nu)]$$

$$\left.-\tfrac12 J_\mu J_\nu+\tfrac14\kappa J_\mu J_\nu+\tfrac16\kappa^2 J_\mu J_\nu+\cdots\right\}.$$

From (10.39), (10.43)–(10.45) we find

$$c(0)=\frac{kTI_1}{3\hbar^2 B}\{(C_\parallel^2+2C_\perp^2)(1-\kappa+\tfrac{13}{6}\kappa^2+\cdots)+(2C_\perp^2+4C_\perp C_\parallel)(\tfrac12\kappa-\tfrac{13}{12}\kappa^2+\cdots)\}$$

$$=\frac{kTI_1}{3\hbar^2 B}(\{C_\parallel^2+2C_\perp^2)-\kappa(C_\perp-C_\parallel)^2+\tfrac{13}{6}\kappa^2(C_\perp-C_\parallel)^2+\cdots\} \quad (10.46)$$

and hence, from (10.27),

$$\frac{1}{T_1}=\frac{1}{T_2}=\frac{2kTI_1}{3\hbar^2 B}\{(C_\parallel^2+2C_\perp^2)-\kappa(C_\perp-C_\parallel)^2+\tfrac{13}{6}\kappa^2(C_\perp-C_\parallel)^2+\cdots\} \quad (10.47)$$

in the extreme narrowing approximation. This agrees with the result of Hubbard (1974, 1981).

To calculate the spin-rotational correlation time from (10.29) we note that using (10.8) and (10.42)

$$\sum_{\mu=1}^{3}\sum_{m=-1}^{1}(-)^m b'_{m\mu}b'_{-m\mu}I_\mu = I_1\sum_{\mu=1}^{3}\sum_{m=-1}^{1}b'_{m\mu}b'^*_{m\mu}$$

$$= I_1(C_\parallel^2 + 2C_\perp^2).$$

Hence from (10.29) and (10.46)

$$\tau_{sr} = \frac{(C_\parallel^2 + 2C_\perp^2) - \kappa(C_\perp - C_\parallel)^2 + \tfrac{13}{6}\kappa^2(C_\perp - C_\parallel)^2 + \cdots}{B(C_\parallel^2 + 2C_\perp^2)}, \qquad (10.48)$$

so that we may express (10.47) as

$$\frac{1}{T_1} = \frac{1}{T_2} = \frac{2I_1 kT(C_\parallel^2 + 2C_\perp^2)\tau_{sr}}{3\hbar^2}. \qquad (10.49)$$

When the spin-rotational tensor of (10.1) is isotropic, its elements $C_{\mu\nu}$ will be a real constant a times the elements of the unit matrix. Since by (B.15) the rotation of a coordinate system produces a unitary transformation and since a unitary transformation leaves the identity operator invariant, it follows for an isotropic tensor that

$$C'_{\mu\nu} = C_{\mu\nu} = a\delta_{\mu\nu}.$$

Hence, from (10.41), $C_\perp$ and $C_\parallel$ are equal and (10.47), (10.48) yield

$$\frac{1}{T_1} = \frac{1}{T_2} = \frac{2kTI_1(C_\parallel^2 + 2C_\perp^2)}{3\hbar^2 B} \qquad (10.50)$$

$$\tau_{sr} = B^{-1}. \qquad (10.51)$$

As was pointed out in subsection 2.4.2, B^{-1} is called the friction time τ_F and it is the correlation time for a component of angular velocity of the sphere. The relations (10.50) and (10.51) may also be deduced by allowing κ to tend to zero in (10.47) and (10.48). Referring to (2.61) and (10.38) we may say that assuming the spin-rotational tensor to be isotropic leads to the rotational diffusion limits of (10.47) and (10.48).

The investigation of relaxation of linear molecules by spin-rotational interactions proceeds along lines similar to those followed for the spherical molecule (McConnell, 1982c). We shall therefore not present the calculations but rather draw attention to some points of difference. We take the origin of the molecular coordinate system at the centre and the third axis along the line of the molecule. In the model the moment of inertia I_3 about the third axis is neglected. Equation (10.34) is replaced by

$$I_1\dot{\omega}_1(t) = -I_1 B\omega_1(t) + A_1(t)$$
$$I_1\dot{\omega}_2(t) = -I_1 B\omega_2(t) + A_2(t),$$

where I_1 is the moment of inertia about a line through the centre and perpendicular to the molecule. Since the angular momentum about the third axis is zero, v in (10.2) and subsequent equations can assume only the values 1 and 2. Then, since $C'_{\mu 3}$ vanishes and since the third axis is an axis of symmetry C_∞, we deduce from (10.41) that $C_\parallel$ vanishes.

Lengthy calculations have been performed to provide second order inertial corrections to results based on the Debye approximation. It was found that the corrections just vanish. Hence the values of the longitudinal and transverse relaxation times in the extreme narrowing approximation and of the spin-rotation correlation time may be deduced from (10.50) and (10.51) by putting $C_\parallel$ equal to zero; that is,

$$\frac{1}{T_1} = \frac{1}{T_2} = \frac{4kTI_1 C_\perp^2 \tau_F}{3\hbar^2} \tag{10.52}$$

$$\tau_{sr} = \tau_F. \tag{10.53}$$

We now examine to what extent the methods of the previous sections are applicable to the asymmetric rotator model of the molecule. Assuming that the components of angular velocity obey the Euler–Langevin equations (2.57) we define the small dimensionless quantity ε by

$$\varepsilon = \frac{(kT)^{1/2}}{(I_1 I_2 I_3 B_1^2 B_2^2 B_3^2)^{1/6}}.$$

Then the components of steady state angular velocity are expanded as series in ε (Ford *et al.*, 1979):

$$\omega_i(t) = \varepsilon\omega_i^{(1)}(t) + \varepsilon^2\omega_i^{(2)}(t) + \varepsilon^3\omega_i^{(3)}(t) + \cdots,$$

where $\varepsilon\omega_i^{(1)}(t)$ is a centred Gaussian variable satisfying

$$\varepsilon^2\langle\omega_i^{(1)}(t)\omega_j^{(1)}(s)\rangle = \frac{\delta_{ij}kT}{I_i}e^{-B_i|t-s|}$$

and $\omega_i(t)$ is centred but non-Gaussian. When we work in three-dimensional representation, we may deduce from (10.40) that J_1^2, J_2^2, J_3^2 commute with each other. Then the value of $\langle R(t)\rangle$ assumes a comparatively simple form given by (McConnell, 1980*b*, section 12.5)

$$\langle R(t)\rangle = \left\{E + \sum_{i=1}^{3}\frac{kT}{I_i B_i^2}(1 - e^{-B_i t}) + \cdots\right\}e^{Gt} \tag{10.54}$$

with

$$G = -\sum_{i=1}^{3} D_i J_i^2, \qquad D_i = D_i^{(1)} + D_i^{(2)}$$

$$D_1^{(1)} = \frac{kT}{I_1 B_1}, \qquad D_2^{(1)} = \frac{kT}{I_2 B_2}, \qquad D_3^{(1)} = \frac{kT}{I_3 B_3} \qquad (10.55)$$

and $D_i^{(2)}$ a quantity of order $\varepsilon^2 D_i^{(1)}$.

The approximation to which $\langle R(t) \rangle$ has been calculated will allow us to employ (10.33) only as far as the term $\varepsilon^2 F^{(2)}(t)$ in the series:

$$\langle R(t)\omega_\mu(t)\omega_\nu(0)\rangle = \langle (E + \varepsilon F^{(1)}(t) + \varepsilon^2 F^{(2)}(t) + \cdots)\omega_\mu(t)\omega_\nu(0)\rangle \langle R(t)\rangle.$$

We express $\omega_i^{(2)}(t)$ as an integral of a product of two $\omega^{(1)}(t)$'s with an exponential factor, and so calculate $\langle \varepsilon F^{(1)}(t)\omega_\mu(t)\omega_\nu(0)\rangle \langle R(t)\rangle$ and $\langle \varepsilon^2 F^{(2)}(t)\omega_\mu(t)\omega_\nu(0)\rangle \langle R(t)\rangle$. We deduce the result (McConnell, 1982a):

$$\int_0^\infty e^{-st} \langle R(t)\omega_\mu(t)\omega_\nu(0)\rangle \, dt$$

$$= \delta_{\mu\nu} E \frac{kT}{I_\mu}\left[(-G + [B_\mu + s]E)^{-1} \right.$$

$$\left. + \sum_{i=1}^{3} \frac{kT}{I_i B_i^2} J_i^2 \{ (-G + [B_\mu + s]E)^{-1} - (-G + [B_\mu + B_i + s]E)^{-1} \} \right]$$

$$+ \delta_{\mu\nu} E \left(\frac{I_\rho - I_\sigma}{I_\mu}\right)^2 \frac{(kT)^2}{I_\rho I_\sigma (B_\rho + B_\sigma - B_\mu)^2} [(-G + [B_\mu + s]E)^{-1}$$

$$- (-G + [B_\rho + B_\sigma + s]E)^{-1} - (B_\rho + B_\sigma - B_\mu)(-G + [B_\mu + s]E)^{-2}]$$

$$- \frac{(kT)^2}{I_1 I_2 I_3} i(\mathbf{J} \cdot \mathbf{e}_\mu \times \mathbf{e}_\nu) \left[(I_\mu - I_r) \left\{ \frac{(-G + [B_\nu + s]E)^{-1}}{(B_\mu - B_\nu)(B_\mu - B_\nu + B_r)} \right. \right.$$

$$+ \frac{(-G + [B_\mu + B_r + s]E)^{-1}}{B_r(B_\mu - B_\nu + B_r)} - \frac{(-G + [B_\mu + s]E)^{-1}}{B_r(B_\mu - B_\nu)} \right\} \qquad (10.56)$$

$$+ (I_\nu - I_r) \left\{ \frac{(-G + [B_\mu + s]E)^{-1}}{(B_\mu - B_\nu)(B_\mu - B_\nu - B_r)} - \frac{(-G + [B_\nu + B_r + s]E)^{-1}}{B_r(B_\mu - B_\nu - B_r)} \right.$$

$$\left. \left. + \frac{(-G + [B_\nu + s]E)^{-1}}{B_r(B_\mu - B_\nu)} \right\} \right]$$

$$- \frac{(kT)^2}{I_\mu I_\nu} i(\mathbf{J} \cdot \mathbf{e}_\mu \times \mathbf{e}_\nu) \left[\frac{(-G + [B_\nu + s]E)^{-1}}{B_\mu(B_\mu - B_\nu)} - \frac{(-G + [B_\mu + s]E)^{-1}}{B_\nu(B_\mu - B_\nu)} \right.$$

$$\left. + \frac{(-G + [B_\mu + B_\nu + s]E)^{-1}}{B_\mu B_\nu} \right]$$

$$-\frac{(kT)^2 J_\mu J_\nu}{I_\mu I_\nu B_\mu B_\nu}\left[(-G+sE)^{-1}-(-G+[B_\mu+s]E)^{-1}\right.$$

$$\left.-(-G+[B_\nu+s]E)^{-1}+(-G+[B_\mu+B_\nu+s]E)^{-1}\right]+\cdots.$$

In the right hand side ρ and σ are chosen such that $\rho\sigma\mu$ is a cyclic permutation of 123, and r is the number which with μ and ν, and $\mu\neq\nu$, constitutes the set 123. The presence of the operator G in (10.56) and the absence of equations like (10.42) would make the calculation of relaxation times very difficult for a molecule with no special symmetry.

Let us therefore study the particular case of a symmetric rotator molecule and examine the situation where the nucleus in which we are interested lies on the axis of symmetry (McConnell, 1982b). Then (10.42) are true. In the approximation employed for the derivation of (10.54), which is to one order beyond that coming from rotational diffusion theory, we may neglect $D_i^{(2)}$ and write (10.55) as

$$G=-(D_1 J_1^2+D_1 J_2^2+D_3 J_3^2), \tag{10.57}$$

where

$$D_1=\frac{kT}{I_1 B_1}, \qquad D_3=\frac{kT}{I_3 B_3}. \tag{10.58}$$

On substituting from (10.40) we deduce from (10.57) that

$$(-G+aE)^{-1}=\begin{bmatrix}(D_1+D_3+a)^{-1} & 0 & 0 \\ 0 & (2D_1+a)^{-1} & 0 \\ 0 & 0 & (D_1+D_3+a)^{-1}\end{bmatrix}. \tag{10.59}$$

We employ (10.26), (10.42), (10.56), (10.58) and (10.59) and put s equal to zero in order to obtain $c(0)$ from

$$c(0)=\frac{1}{3\hbar^2}\sum_{\mu,\nu=1}^{3}I_\mu I_\nu\sum_{m,n=-1}^{1}(-)^m b'_{n\mu}b'_{m\nu}\left(\int_0^\infty \langle R(t)\omega_\mu(t)\omega_\nu(0)\rangle\,dt\right)_{n,-m}. \tag{10.60}$$

We find that the $J_\mu J_\nu$-terms give no contribution. An elementary calculation yields the final result

$$c(0)=\frac{kT}{3\hbar^2}\left\{\frac{2I_1 C_\perp^2}{B_1+D_1+D_3}+\frac{I_3 C_\parallel^2}{B_3+2D_1}\right.$$

$$+kT\left[\left(\frac{2}{B_1^3}+\frac{2I_1}{I_3 B_1^2 B_3}-\frac{2I_3}{I_1 B_1^2(B_1+B_3)}\right)C_\perp^2\right.$$

$$\left.\left.+\frac{2I_3(C_\parallel^2+2C_\perp C_\parallel)}{I_1 B_1 B_3(B_1+B_3)}\right]\right\}. \tag{10.61}$$

In the extreme narrowing approximation (10.27) and (10.61) give

$$\frac{1}{T_1} = \frac{1}{T_2} = \frac{2kT}{3\hbar^2}\left\{\frac{2I_1 C_\perp^2}{B_1 + D_1 + D_3} + \frac{I_3 C_\parallel^2}{B_3 + 2D_1}\right.$$

$$+ kT\left[\left(\frac{2}{B_1^3} + \frac{2I_1}{I_3 B_1^2 B_3} - \frac{2I_3}{I_1 B_1^2(B_1 + B_3)}\right)C_\perp^2\right.$$

$$\left.\left.+ \frac{2I_3(C_\parallel^2 + 2C_\perp C_\parallel)}{I_1 B_1 B_3(B_1 + B_3)}\right]\right\}. \tag{10.62}$$

By using (10.26) in place of (10.60) we can easily calculate the relaxation rates outside the extreme narrowing approximation. We find from (10.42) that

$$\sum_{\mu=1}^{3}\sum_{m=-1}^{1}(-)^m b'_{-m\mu} b'_{m\mu} I_\mu = \sum_{\mu=1}^{3}\sum_{m=-1}^{1}|b'_{m\mu}|^2 I_\mu$$

$$= 2I_1 C_\perp^2 + I_3 C_\parallel^2.$$

Hence, from (10.29) and (10.61),

$$\tau_{\text{sr}} = \frac{1}{2I_1 C_\perp^2 + I_3 C_\parallel^2}\left\{\frac{2I_1 C_\perp^2}{B_1 + D_1 + D_3} + \frac{I_3 C_\parallel^2}{B_3 + 2D_1}\right.$$

$$+ kT\left[\left(\frac{2}{B_1^3} + \frac{2I_1}{I_3 B_1^2 B_3} - \frac{2I_3}{I_1 B_1^2(B_1 + B_3)}\right)C_\perp^2\right.$$

$$\left.\left.+ \frac{2I_3(C_\parallel^2 + 2C_\perp C_\parallel)}{I_1 B_1 B_3(B_1 + B_3)}\right]\right\}. \tag{10.63}$$

To see what (10.62) and (10.63) become in the rotational diffusion limit we note that (2.61) combined with (10.58) makes D_1/B_1, D_1/B_3, D_3/B_1, D_3/B_3 vanishingly small. Consequently (10.62) become

$$\frac{1}{T_1} = \frac{1}{T_2} = \frac{2kT}{3\hbar^2}\left\{\frac{2I_1 C_\perp^2}{B_1} + \frac{I_3 C_\parallel^2}{B_3}\right\}, \tag{10.64}$$

and (10.63) becomes

$$\tau_{\text{sr}} = \frac{1}{2I_1 C_\perp^2 + I_3 C_\parallel^2}\left\{\frac{2I_1 C_\perp^2}{B_1} + \frac{I_3 C_\parallel^2}{B_3}\right\}. \tag{10.65}$$

11

Theory and experiment for relaxation processes

11.1 Introductory remarks

In this final chapter an attempt will be made to relate the theory of nuclear magnetic relaxation developed in Chapters 3–10 with the results of a limited number of the numerous relaxation experiments described in the literature. We have ignored internal molecular motions but have considered both translational and rotational motion of the molecules, when this was appropriate. The theory that we have elaborated allows us in principle to discuss molecules that have no special symmetry, but in practice our explicit results were obtained only for spherical, linear, symmetric rotator and planar molecules. Similarly in order to derive manageable analytical results we have adopted certain simplifying assumptions for the chemical shielding and the spin-rotation tensor. Moreover, while our calculations have usually been based on Brownian motion theory in which inertial results are included, we have frequently employed the rotational diffusion limits of the results of the inertial calculations. Another simplification that has often been introduced is the restriction of results to the extreme narrowing approximation.

For nuclear magnetic relaxation the dominant mechanism is usually dipole–dipole interaction, be it intermolecular or intramolecular. However, in section 7.2 it was shown that in certain circumstances scalar interaction by chemical exchange could be a relaxation process more important than that arising from intramolecular dipolar coupling. We saw in section 9.1 that quadrupolar interaction may exist when the spin of the interacting nucleus is greater than $\frac{1}{2}$. Then in fact the associated relaxation effect may greatly exceed that resulting from dipolar interaction. Relaxation by anisotropic chemical shift was found to increase with the Larmor frequency and consequently with the external field. No such comparison was feasible for spin-rotational interaction, since the complexity of the calculation allowed us to give explicit results only in the extreme narrowing approximation.

With the possible exception of intermolecular interactions all the relaxation processes result from random rotational motion. If the molecules under consideration are spherical, if each molecule contains only one nucleus and if it is at the centre, the relaxation is due entirely to random translational motion; otherwise it is due to both rotational and translational motions as we saw, for example, in section 5.4. In the examination of experimental results one may have to consider the relative importance of several competing relaxation processes. The mathematical discussion of this question may be complicated by the presence of tensors in the interaction Hamiltonians, as we see from (8.17) for anisotropic chemical shift and from (9.19) and (9.20) for quadrupole interactions. The respective tensors will have different principal axes and simplifications may be needed to obtain perspicuous solutions.

When both intermolecular and intramolecular interactions are responsible for a relaxation process in a pure liquid, the relaxation rate due to each interaction may often be found experimentally by employing a *dilution technique*. The pure liquid is mixed with another whose molecules have nuclei which are isotopic to those of the original liquid but which have magnetic moments sufficiently small that the dipolar interactions between the different types of nuclei and between the new nuclei are negligible. If then we extrapolate the experimental results up to what would happen for infinite dilution, we are left with the relaxation due to intramolecular interactions. On subtracting this from the measured total relaxation rate there remains the relaxation due to intermolecular interactions only.

The dilution technique is frequently applied to substances whose molecules contain protons H. In the dilution process one or more of these protons is replaced by a deuteron D; when all the protons have been so replaced, the substance is said to be perdeuterated. Let us employ the results of Chapter 6 to make a rough comparison between the relative strengths of H–H, H–D and D–D dipolar interactions. The magnetic moments μ_H of the proton and μ_D of the deuteron are given by (Pople *et al.*, 1959, Appendix A)

$$\mu_H = 2.79 \frac{eh}{2Mc}, \qquad \mu_D = 0.86 \frac{eh}{2Mc}, \qquad (11.1)$$

where M is the proton mass. Since the spin of H is $\frac{1}{2}$ and the spin of D is unity, the relation $\mu = \gamma h \mathbf{I}$ and (11.1) yield

$$\gamma_H = \frac{2 \times 2.79 e}{2Mc}, \qquad \gamma_D = \frac{0.86 e}{2Mc} \qquad (11.2)$$

for the respective gyromagnetic ratios, and so

$$\frac{\gamma_H}{\gamma_D} = 6.49. \qquad (11.3)$$

We shall endeavour to obtain an estimate of the relative relaxation rates for the H–H, H–D, D–D systems. Our calculations will apply to both intermolecular and intramolecular interactions. With an obvious notation we see from (4.20) that

$$\frac{j_{HH}(\omega)}{j_{HD}(\omega)} = \frac{j_{HD}(\omega)}{j_{DD}(\omega)} = \frac{\gamma_H^2}{\gamma_D^2}. \tag{11.4}$$

Working in the extreme narrowing approximation we deduce from (4.60) and (11.4) that

$$\left(\frac{1}{T_1}\right)_{HH} = \frac{3}{8}\frac{\gamma_H^4}{\gamma_D^4}\left(\frac{1}{T_1}\right)_{DD}. \tag{11.5}$$

In the case of H–D we know that there is no single longitudinal relaxation time but as an estimate of it we take T_1''. Then we have from (4.90) and (4.91)

$$(T_1^{-1})_{HH} = 5j_{HH}(0), \qquad (T_1^{-1})_{HD} = \tfrac{80}{9}j_{HD}(0),$$

so that

$$(T_1^{-1})_{HH} = \frac{9}{16}\frac{\gamma_H^2}{\gamma_D^2}(T_1^{-1})_{HD}. \tag{11.6}$$

Employing (11.3) we deduce from (11.5) and (11.6)

$$(T_1^{-1})_{HH} = 23.7(T_1^{-1})_{HD} \tag{11.7}$$

$$(T_1^{-1})_{HH} = 665(T_1^{-1})_{DD}. \tag{11.8}$$

Equations (11.7) and (11.8) show that the dipolar relaxation rate for H–D is much smaller than that for H–H, as is required for the use of the dilution technique, and that the influence of the D–D interactions on the dipole relaxation process is entirely negligible.

Suppose now that we have a dilution experiment where liquid molecules containing protons are mixed with the corresponding perdeuterated liquid. Let α be the fraction of the mixture constituted by the original liquid, so that $1-\alpha$ is the fraction constituted by the perdeuterated liquid. We concern ourselves only with dipolar interactions. Thus the total longitudinal relaxation rate is the sum of the dipolar intramolecular and intermolecular rates. The deuteron plays no part in the intramolecular interaction. For the intermolecular rate a fraction α has the H–H rate and the fraction $1-\alpha$ has the H–D rate, which is expressed by (11.6) and (11.7) in terms of the intermolecular H–H rate. Hence the observed rate

$$\frac{1}{T_1} = \left(\frac{1}{T_1}\right)_{intra} + \left\{\alpha + \frac{16(1-\alpha)\gamma_D^2}{9\gamma_H^2}\right\}\left(\frac{1}{T_1}\right)_{inter}; \tag{11.9}$$

that is

$$\frac{1}{T_1} = \left(\frac{1}{T_1}\right)_{intra} + \{0.042 + 0.958\alpha\}\left(\frac{1}{T_1}\right)_{inter} \tag{11.10}$$

Equations (11.9) and (11.10) express the spin–lattice relaxation time of the mixture in terms of the intermolecular and intramolecular spin–lattice relaxation times for the non-deuterated liquid.

It has been found that serious errors were made in interpreting the results of early experiments on relaxation rates, since no account was taken of the presence of dissolved gaseous oxygen in the sample under examination. The oxygen is paramagnetic and to study the influence of dissolved paramagnetic ions on T_1 and T_2 in diamagnetic liquids we employ (4.78), where I refers to the nuclei in the sample and S to the electronic spin of the paramagnetic ions. The electrons are subject to electron spin resonance phenomena including relaxation (Atherton, 1973). Since the spin magnetic moment of the electron is $eh/(2m_e c)$, where m_e is the mass of the electron, and its spin is $\frac{1}{2}$, it follows that its gyromagnetic ratio is $e/(m_e c)$. The order of this is 10^3 times that of a nuclear gyromagnetic ratio, as we see from (11.2). Hence the electron spin relaxation time is several orders of magnitude smaller than the nuclear relaxation time. Consequently on denoting the electron spin by S we may put $\langle S_z \rangle$ equal to $\langle S_z \rangle_0$ in (4.78), which then becomes

$$\frac{d(\langle I_z \rangle - \langle I_z \rangle_0)}{dt} = -\frac{\langle I_z \rangle - \langle I_z \rangle_0}{T_1^{II}},$$

and so there is a single longitudinal relaxation time T_1^{II}. In the extreme narrowing approximation (4.81) yields

$$\frac{1}{T_1^{II}} = \tfrac{40}{9} S(S+1) j(0) = \tfrac{10}{3} j(0)$$

and, since in analogy with (11.4)

$$\frac{j_{He}(\omega)}{j_{HH}(\omega)} = \frac{\gamma_e^2}{\gamma_H^2},$$

we see that the order of $j(0)$ is 10^6 that of the spectral density for H–H. Thus the presence of an extremely small proportion of oxygen in the liquid can alter the relaxation rate considerably.

Various means have been devised to remove gaseous oxygen from a sample. One such means is to deposit mischmetal on the inner surface of the container of the sample (Rugheimer & Hubbard, 1963). The oxygen is then adsorbed on the mischmetal. Another way of removing the oxygen is by freezing the sample, evacuating the gas and thawing. On repeating this operation many times the sample may become oxygen-free (Bonera & Rigamonti, 1965a).

In the interpretation of experimental results for nuclear magnetic relaxation frequent use is made of the Rayleigh–Stokes macroscopic laws of friction

$$\zeta_t = 6\pi a \eta \tag{11.11}$$

$$\zeta_r = 8\pi a^3 \eta. \tag{11.12}$$

These provide the coefficients of translational friction ζ_t and of rotational friction ζ_r for a sphere of radius a in a medium with viscosity coefficient η. Equations (11.11) and (11.12) were employed in section 6.6 for the discussion of the conditions for extreme narrowing in translational and rotational Brownian motion, and it was found that these conditions were essentially equivalent, if one assumed the relation

$$\zeta_r = \frac{4a^3}{3} \zeta_t \tag{11.13}$$

obtained by eliminating η between (11.11) and (11.12). Rugheimer & Hubbard (1963) made a theoretical and experimental study of the relaxation of the protons in liquid methane CH_4 for temperatures between -182 and $-164°C$. They considered contributions to the relaxation rate from spin-rotational, intramolecular dipole–dipole and intermolecular dipole–dipole interactions. They found that agreement between theory and experiment is poor, if one accepted (11.13), but that exact agreement could be obtained if (11.13) were replaced by

$$\zeta_r = \frac{4a^2}{3c} \zeta_t \tag{11.14}$$

and the parameter c were chosen for each temperature.

This question was later studied by Zeidler (1965) in relation to measurements of the spin–lattice relaxation time of protons and deuterons in ten organic liquids at 25°C. For the microscopic domain he replaced (11.11) and (11.12) by

$$\zeta_t = 6\pi a \eta f_t \tag{11.15}$$

$$\zeta_r = 8\pi a^3 \eta f_r, \tag{11.16}$$

where η is the macroscopic viscosity and f_t, f_r are translational and rotational microscopic factors. We see from (11.14)–(11.16) that $c = f_t/f_r$, and we deduce from Zeidler's Table 3 that f_t/f_r lies between 3 and 8 for the ten liquids. This shows that an assumption that the same η is present in (11.11) and (11.12) is unjustifiable. The value of f_t in (11.15) varies between 0.46 and 0.79, and the value of f_r in (11.16) varies between 0.06 and 0.17 for the different liquids. Hence (11.12) is incorrect by an order of magnitude. It therefore appears unwise to base a nuclear magnetic relaxation theory on the Rayleigh–Stokes laws.

11.2 Summary of theoretical results

In order to form a general picture of the implications for experiment of theoretical results derived in previous chapters we collect in Table 11.1 expressions for relaxation and correlation times from sections 5.2, 5.3, 6.2, 6.7, 7.2, 7.3, 8.3, 9.3, 10.4 and Appendix D. Since we are interested here in general features, we give results only for the spherical molecular model. For non-spherical models information may be obtained by starting from the references below the table. When employing rotational Brownian motion theory we present only rotational diffusion results; the results of inertial theory may easily be traced from the references. We have found in sections 6.3, 6.5, 8.3 and 9.3 that expressions for relaxation times obtained in rotational diffusion theory for spherical molecular models may be applicable to linear and symmetric rotator models but not to a circular plate model, as we can immediately verify by comparing (6.40) with (6.13). For the reason given in section 5.3 no attempt has been made to give an inertial Brownian motion theory of relaxation by intermolecular dipolar interaction. Finally in Table 11.1 the relaxation times are presented only for the extreme narrowing approximation but again the fuller results are easily accessible in the previous chapters. The significance of the symbols used in the table may be learned by referring to the above sections.

Table 11.1 shows that there are three distinct correlation times for the rotational diffusion theory of spherical molecules, namely the chemical exchange time τ_e which for liquids is of order 10^{-6} s, the time $\zeta_r/(6kT)$ or $I_1 B/(6kT)$ of order 10^{-11} s and τ_F with value B^{-1} of order 10^{-13} s. For all interactions the correlation time τ'_j is defined as that corresponding to the function $H_q(t)$ in the $G(t)$ of (3.38), namely

$$\sum_{q=-j}^{j} (-)^q H_{-q}(t) A^{(q)}$$

with $j=0$ for scalar interaction, $j=1$ for spin-rotational interaction and $j=2$ for the other relaxation mechanisms. Since the random rotational motion is described by $H_q(t)$, this definition of correlation time is a reasonable one. Apart from the case of spin-rotational interaction, for which $H_q(t)$ is given by (10.15) as

$$\hbar^{-1} \sum_{v=1}^{3} \sum_{m=-1}^{1} b'^*_{mv} I_v D^{1*}_{qm}(\alpha(t), \beta(t), \gamma(t)) \omega_v(t), \qquad (11.17)$$

$H_q(t)$ depends only on angular variables. It is the presence of the angular velocity component $\omega_v(t)$ in (11.17) that produces a correlation time for spin-rotational interaction much smaller than that for the other relaxation mechanisms.

Table 11.1 *Relaxation times and interaction correlation times for relaxation processes*

Relaxation interaction	Relaxation times	Correlation time
Intermolecular dipolar		
random walk[a]	$\dfrac{1}{T_1} = \dfrac{1}{T_2} = \dfrac{8\pi N\gamma^4\hbar^2 I(I+1)}{15D'd_0}\left\{1+\dfrac{5}{12}\dfrac{\langle r^2\rangle}{d_0^2}\right\}$	
Brownian motion[b]	$\dfrac{1}{T_1} = \dfrac{1}{T_2} = \dfrac{8\pi N\gamma^4\hbar^2 I(I+1)}{15d_0 kT}\zeta_r$	$\dfrac{\zeta_r}{6kT}$
Intramolecular dipolar		
like spins[c]	$\dfrac{1}{T_1} = \dfrac{1}{T_2} = \dfrac{\gamma^4\hbar^2 I(I+1)}{3r^6 kT}\zeta_r$	
unlike spins[d]	$\dfrac{1}{T_1^{II}} = \dfrac{2}{T_1^{SI}} \qquad \dfrac{1}{T_1^{II}} = \dfrac{1}{T_2^{SI}} = \dfrac{2\gamma_I^2\gamma_S^2\hbar^2 S(S+1)}{9r^6 kT}\zeta_r$	$\dfrac{\zeta_r}{6kT}$
Scalar		
chemical exchange[e]	$\dfrac{1}{T_1^{II}} = -\dfrac{1}{T_1^{SI}} = \dfrac{2}{3}A^2 S(S+1)\tau_e$ $\dfrac{1}{T_2^{II}} = \dfrac{1}{3}A^2 S(S+1)\left\{\tau_e+\dfrac{\tau_e}{1+(\omega_I-\omega_S)^2\tau_e^2}\right\}$	τ_e
second kind[f]	$\dfrac{1}{T_1} = \dfrac{\frac{2}{3}A^2 S(S+1)\tau_T}{1+(\omega_I-\omega_S)^2\tau_T^2}$ $\dfrac{1}{T_2} = \dfrac{1}{3}A^2 S(S+1)\left\{\tau_L+\dfrac{\tau_T}{1+(\omega_I-\omega_S)^2\tau_T^2}\right\}$	τ_T
Chemical shift[g]	$\dfrac{1}{T_1} = \dfrac{6}{7T_2} = \dfrac{1}{20}\gamma^2 H_0^2(1+\tfrac{1}{3}\zeta^2)\dfrac{\delta_z^2\zeta_r}{kT}$	$\dfrac{\zeta_r}{6kT}$
Quadrupole[h]	$\dfrac{1}{T_1} = \dfrac{1}{T_2} = \dfrac{1}{80}(1+\tfrac{1}{3}\eta^2)\left(\dfrac{e^2 qQ}{h}\right)^2\dfrac{2I+3}{I^2(2I-1)}\dfrac{\zeta_r}{kT}$	$\dfrac{\zeta_r}{6kT}$
Spin-rotation[i]	$\dfrac{1}{T_1} = \dfrac{1}{T_2} = \dfrac{2(C_\parallel^2+2C_\perp^2)I_1\tau_F}{3\hbar^2}kT$	τ_F

[a] (5.24); [b] (5.30); [c] (6.13), (D.31); [d] (6.61)–(6.63); [e] (7.6), (7.8); [f] (7.22); [g] (8.40); [h] (9.46); [i] (10.50), (10.51).

We see from (4.19) that for intramolecular dipolar interactions $H_q(t)$ is proportional to $Y_{2q}(\theta(t), \phi(t))$ and so for all molecular models τ_2' is equal to the correlation time for Y_{2q}, which we denoted by τ_2. Then the correlation time is in fact independent of the relaxation process. According to sections 8.1 and 9.3 τ_2' is equal to τ_2 for relaxation by anisotropic chemical shift and by quadrupolar interaction only if the respective asymmetry parameter ζ or η vanishes. In the last column of Table 11.1 it is assumed that ζ and η vanish. However, caution must be exercised before using a value of τ_2' known experimentally for one relaxation process in calculations for another process. The inertial theory yields small corrections of relative order $kT/(I_1 B^2)$ to the rotational diffusion values of the correlation times. As we saw in (6.26), there may exist an infinity of values of correlation times τ_2 for the same molecular model.

The constant external field H_0 enters explicitly into the expressions for relaxation rates only in the case of anisotropic chemical shift: it enters implicitly through ω_I and ω_S for scalar interaction, these angular frequencies being retained in the extreme narrowing approximation for the reasons explained in section 7.2.

To acquire a knowledge of the experimental literature on nuclear magnetic relaxation one may consult the book of Abragam (1961) and the review articles of Hertz (1967, 1983) and of Farrar, Maryott & Malmberg (1972). It should be remembered that agreement between a theoretical and experimental result obtained for one liquid will not necessarily be true for other liquids, whose molecules have a similar shape. Thus while Rugheimer & Hubbard (1963) found that the relation (11.13) does not hold for liquid methane CH_4, they found in the same set of experiments that it does hold for liquid carbon tetrafluoride CF_4. Indeed to a large extent one has to examine each liquid on its own, and, by comparing the experimental relaxation rates with those coming from the theory, endeavour to simultaneously make a judgement on the reliability of existing theories and to identify the competing relaxation mechanisms that could have a significant influence on the observed results.

In the following sections we give examples of how results derived in previous chapters have been applied to the interpretation of experiments on nuclear magnetic relaxation. The next section is devoted to intermolecular dipolar relaxation and the following ones are concerned with intramolecular relaxation processes.

11.3 Intermolecular dipole–dipole interactions

For the hard sphere model of Torrey (1953) we have from (5.24) the longitudinal intermolecular dipole–dipole relaxation rate

$$\frac{1}{T_1} = \frac{8\pi\gamma^4\hbar^2 N I(I+1)}{15 D' d_0}\left\{1 + \frac{5}{12}\frac{\langle r^2\rangle}{d_0^2}\right\}. \tag{11.18}$$

For this N is the number of spins per unit volume, d_0 is the distance of closest approach, τ is the mean time between flights called the *jump time*, $\langle r^2\rangle$ is the mean square distance between flights, $\langle r^2\rangle^{1/2}$ being called the *jump length*, and D' is the self-diffusion coefficient defined in (2.43) as $\langle r^2\rangle/6\tau$.

Bender & Zeidler (1971) performed experiments on dipolar proton–proton relaxation in liquid chloroform. The molecule $CHCl_3$ is approximately spherical, and it obviously has no intramolecular proton–proton interactions. The value of T_1^{-1} was measured in neat chloroform and in $CHCl_3/CDCl_3$ mixtures in the temperature range $-18°C$ to $47°C$. Putting $I=\frac{1}{2}$ in (11.18) we obtain

$$\frac{1}{T_1}\frac{D'}{N} = \pi\gamma^4\hbar^2\left\{\frac{2}{5d_0} + \frac{\langle r^2\rangle}{6d_0^3}\right\}, \tag{11.19}$$

which is temperature independent if $\langle r^2\rangle$ is so. N is now interpreted as the number of protons per cubic centimetre. It is assumed in the Torrey model that the interacting nuclei are at the centres of the hard spheres, and d_0, being twice the molecular radius, is here estimated as lying between 0.41 and 0.56 nm. The value of D' was measured by spin-echo technique, and it was deduced that $D'/T_1 N$ is approximately independent of the temperature. Assigning to it the value $3 \times 10^{-39}\,\mathrm{m^5\,s^{-2}}$ they calculated that the jump length has the constant value 55 nm. From this and (2.43) it follows that the jump time increases from 1.6×10^{-11} s at $47°C$ to 3.9×10^{-11} s at $-18°C$. It appears that these figures are too small when they are compared with results of experiments on cyclopentane (De Graaf, 1969) and *n*-propanol (Larsson, 1968).

One possible explanation of this discrepancy is the inadequacy of the hard sphere model, as was noted in section 5.4. Another is that (2.38) and therefore (11.19) were derived under the assumption that the two interacting protons move independently. On account of its r^{-3}-dependence the dipolar interaction is most effective at short distances. A consequence of this is that the $2t$ in (2.38) is altered. By comparing (2.45) with (2.42) we see that the $2t$ produced a relative self-diffusion coefficient $2D'$. Hence, when $2t$ is changed, so also will be the relative self-diffusion coefficient. Such a change has also been proposed by Zeidler (1971) to explain the discrepancies in the values of τ for quasielastic neutron scattering and NMR relaxation. As a result of relaxation measurements by spin-echo technique on crown ether Richter & Zeidler (1985) confirmed that the relative self-diffusion coefficient is not $2D'$.

They also concluded that theories with the dipole at the centre of a spherical molecule yield too small molecular diameters and too large relative diffusion coefficients.

It is therefore seen that the Torrey theory as expressible by (11.18) provides a qualitative but not a quantitative theory of relaxation by intermolecular interaction. Further discussion of the shortcomings of such theories is to be found in Hertz (1967).

As an illustration of the application of Brownian motion theory to intermolecular dipolar interactions we consider the experiments of Albrand *et al.* (1981), who measured the relaxation time T_1 for the ^{13}C nuclei in neopentane $C(CH_3)_4$, whose molecules are approximately spherical with one carbon nucleus at the centre. In order to isolate the dipole–dipole intermolecular interactions ditertiobutyl nitroxide $[(CH_3)_3C]_2NO$ free radicals were introduced with concentrations varying up to $5.6 \times 10^{26} \, m^{-3}$. The dominant relaxation mechanism was then the interactions between the nuclear spin of ^{13}C and the electronic spins of the free radicals. The experiments showed quite clearly that for a given concentration the relaxation time T_1 is different for the central and the off-centre carbon nuclei. The theoretical discussion was based on equations equivalent to (4.21) and (4.81). Two distinct models were chosen, namely a uniform distribution of interacting molecules and a distribution based on a pair correlation function. The latter model provided better agreement with experiment, and it has the advantage that it contains no adjustable parameter. These investigations were later extended to embrace a large range of frequencies (Albrand *et al.*, 1983).

11.4 Intramolecular dipolar and spin-rotational interactions

The chloroform molecule examined by Bender & Zeidler (1971) is also a convenient subject for the study of intramolecular interactions. Let us consider the relaxation of the proton in $CHCl_3$. Since the proton has spin $\frac{1}{2}$, it does not relax by quadrupole interaction. Then, since ^{13}C constitutes only 1.1% of the natural abundance of carbon (Townes & Schawlow, 1975, p. 644) and ^{12}C has zero magnetic moment, the H–C dipolar interaction may be neglected. We shall suppose that the external uniform field is not too strong, so that relaxation by anisotropic chemical shift is negligible. We shall also suppose that there is no relaxation by scalar interaction. The only remaining relaxation mechanisms are H–Cl dipolar interactions and spin-rotational interactions, and we write

$$\frac{1}{T_1} = \left(\frac{1}{T_1}\right)_{\text{H–Cl}} + \left(\frac{1}{T_1}\right)_{\text{sr}}. \tag{11.20}$$

In the preceding section the chloroform molecule was assumed to be spherical in order that experimental results could be compared with the implications of the Torrey theory. We shall now employ Brownian motion theory and so relax the assumption by taking the molecule to be a symmetric top with the C–H bond as the symmetry axis. In dealing with (4.78) for unlike spins in a situation, where it is known from experiment that an exponential relaxation exists, it is frequently presumed that the last term in (4.78) is negligible, so that $T_1 = T_1''$. Thus, in the present case we employ for the extreme narrowing approximation the relaxation rate given for rotational diffusion theory by (6.48) and (6.61), where θ is the angle between the axis of symmetry and the line joining H to Cl. Summing over the three chlorine atoms we deduce that

$$\left(\frac{1}{T_1}\right)_{\text{H–Cl}} = \tfrac{1}{3}\gamma_{\text{H}}^2\overline{\gamma_{\text{Cl}}^2}I_{\text{Cl}}(I_{\text{Cl}}+1)\sum_{i=1}^{3} r_i^{-6}$$

$$\times \left\{\frac{(3\cos^2\theta-1)^2}{6D_1} + \frac{12\sin^2\theta\cos^2\theta}{5D_1+D_3} + \frac{3\sin^4\theta}{2D_1+4D_3}\right\}, \tag{11.21}$$

where $\overline{\gamma_{\text{Cl}}^2}$ denotes the average of γ_{Cl}^2 over ^{35}Cl and ^{37}Cl, the value of I_{Cl} in both cases being 3/2, and r_i is the distance between H and Cl. If we compare (6.51) with (6.52), we see that (11.21) is expressible as

$$\left(\frac{1}{T_1}\right)_{\text{H–Cl}} = \tfrac{2}{9}\gamma_{\text{H}}^2\overline{\gamma_{\text{Cl}}^2}\hbar^2 I_{\text{Cl}}(I_{\text{Cl}}+1)\sum_{i=1}^{3} r_i^{-6}$$

$$\times \left\{1+\frac{3(D_1-D_3)}{5D_1+D_3}\sin^2\theta\left(1+\frac{3(D_1-D_3)}{2(D_1+2D_3)}\sin^2\theta\right)\right\}, \tag{11.22}$$

which agrees with the result of Bender & Zeidler. The values of D_1 and D_3 for the temperature range -35 to $50°$C were found by Huntress (1969). On substituting them and the values of the other constants involved in (11.22), and subtracting the right hand side from the total experimental intramolecular rate Bender & Zeidler deduced from (11.20) the values of the spin-rotational relaxation rate for temperatures between -18 and $47°$C. The dipolar H–Cl rate is approximately constant at about $2.4\times10^{-3}\,\text{s}^{-1}$ while the spin-rotation rate increases with increasing temperature from 1.3×10^{-3} to $1.9\times10^{-3}\,\text{s}^{-1}$.

Since the proton lies on the axis of symmetry of the chloroform molecule, we may apply the theoretical results of Chapter 10 to its relaxation by spin-rotation interactions. In the extreme narrowing approximation the value of

T_1 is given by (10.62). This value differs from that given by the rotational diffusion result (10.64) by a few per cent at most and, since this is well within the limits of experimental error, we shall employ (10.64) when discussing experiments. Introducing the diffusion coefficients from (10.58) we express (10.64) as

$$\left(\frac{1}{T_1}\right)_{sr} = \frac{2}{3}\left\{2D_1\left(\frac{I_1 C_\perp}{\hbar}\right)^2 + D_3\left(\frac{I_3 C_\parallel}{\hbar}\right)^2\right\}. \qquad (11.23)$$

While $C_\parallel$ and $C_\perp$ are found experimentally to be time dependent, they are not strongly so and it is therefore possible to obtain the values of $C_\parallel^2$ and $C_\perp^2$ by employing (11.23) for two temperatures that are close together. Since the chloroform molecule is nearly spherical, we approximate I_3 by I_1 and find I_1 in terms of the rotational constant B_0 for the ground vibrational state from

$$I_1 = \frac{\hbar}{4\pi B_0} \qquad (11.24)$$

(Herzberg, 1966, p. 670). On substituting $B_0 = 3302 \times 10^6$ s^{-1} (Townes & Schawlow, 1975, p. 617) and $\hbar = 1.0546 \times 10^{-34}$ J s we deduce from (11.24) that

$$I_1 = 2.542 \times 10^{-45} \text{ kg m}^2. \qquad (11.25)$$

We have from Huntress (1969) that at $-10°$C temperature $D_1 = 0.71 \times 10^{11}$ s^{-1} and $D_3 = 1.48 \times 10^{11}$ s^{-1} and from Bender & Zeidler that the spin-rotational relaxation rate is 1.5×10^{-3} s^{-1}. Combining these results with (11.23) and (11.25) we find that

$$|C_\parallel| = 4.6 \text{ kHz}; \qquad |C_\perp| = 2.24 \text{ kHz}. \qquad (11.26)$$

Similarly we have for $47°$C that $D_1 = 1.25 \times 10^{11}$ s^{-1}, $D_3 = 1.95 \times 10^{11}$ s^{-1}, $(T_1^{-1})_{sr} = 1.9 \times 10^{-3}$ s^{-1} and

$$|C_\parallel| = 4.6 \text{ kHz}; \qquad |C_\perp| = 1.78 \text{ kHz}. \qquad (11.27)$$

11.5 Anisotropic chemical shift and spin-rotational interaction

It was pointed out in section 8.1 that McConnell & Holm (1956) were the first to propose anisotropic chemical shift as a relaxation mechanism. For the Larmor frequency ν_0 they obtained the spin–lattice relaxation rate for nuclei of spin $\frac{1}{2}$

$$\frac{1}{T_1} = \frac{8\pi^2}{15} \frac{(\Delta\sigma)^2 \nu_0^2 \tau_c}{1 + 4\pi^2 \nu_0^2 \tau_c^2}. \qquad (11.28)$$

If we identify γH_0 with $2\pi\nu_0$ and τ_c with τ_2 defined in (D.31) and if we employ (8.60), then (11.28) is given by (8.54) for the spherical, linear and symmetric rotator models of the molecule, when the asymmetry parameter ζ and the angle β that appears in (8.41) and later equations vanish. According to (11.28) the relaxation time decreases as v_0 increases. On the other hand (6.11) shows that T_1 increases with v_0. McConnell & Holm performed experiments on carbon disulphide $^{13}CS_2$ and carbon tetrachloride $^{13}CCl_4$. On account of the natural scarcity of ^{13}C mentioned at the beginning of the previous section, relaxation of ^{13}C by dipolar interaction is extremely small and would not account for the short relaxation time 60 s of ^{13}C in $^{13}CS_2$, which was therefore ascribed to anisotropic chemical shift.

An interesting comparison between anisotropic chemical shift and spin-rotational relaxation effects was made by Spiess *et al.* (1971), who studied experimentally the relaxation of the ^{13}C nucleus in the linear carbon disulphide molecule $^{13}CS_2$. We have from (8.61), when the anisotropic chemical shift tensor has an axis of cylindrical symmetry that coincides with the line of the molecule and in the extreme narrowing approximation, the relaxation rate

$$\left(\frac{1}{T_1}\right)_{cs} = \frac{4\pi^2 v_0^2 I_1 (\Delta\sigma)^2}{45k T \tau_F}.\tag{11.29}$$

For spin-rotational relaxation we have from (10.52) in the extreme narrowing approximation the rate

$$\left(\frac{1}{T_1}\right)_{sr} = \frac{4kT I_1 C_\perp^2 \tau_F}{3\hbar^2}.\tag{11.30}$$

From (11.29) and (11.30) we deduce that

$$\left(\frac{1}{T_1}\right)_{cs}\left(\frac{1}{T_1}\right)_{sr} = \frac{16\pi^2 v_0^2 I_1^2 (\Delta\sigma)^2 C_\perp^2}{135\hbar^2}.\tag{11.31}$$

For a field corresponding to a Larmor frequency 62 MHz it was found experimentally that the left hand side of (11.31) was roughly constant for temperatures in the range -100 to $40°C$. This indicates that $\Delta\sigma C_\perp$ is constant in this temperature range. For the same field it was found experimentally that the spin-rotation relaxation rate varies linearly with T for the temperature range -100 to $0°C$.

11.6 Quadrupole relaxation

We saw in section 9.1 that nuclear magnetic relaxation by quadrupole interaction may occur for nuclei whose spin exceeds $\frac{1}{2}$ and that, when it does occur, the relaxation rate may be several orders of magnitude greater than those arising from intramolecular dipolar interactions and anisotropic chemical shift. In section 9.2 it was shown that spin–lattice and spin–spin relaxation times exist in the extreme narrowing approximation for $I \geqslant 1$ and outside this approximation for $I = 1$. An important example of a nucleus with spin one is the deuteron, which has an electric quadrupole moment eQ with Q equal to 2.77×10^{-31} m^2 (Pople *et al.*, 1959, Appendix A).

The result of Chapter 9 that has the greatest relevance for experiment is (9.50):

$$\frac{1}{T_1} = \frac{1}{T_2} = \frac{3}{40}\left(\frac{e^2qQ}{\hbar}\right)^2 \frac{2I+3}{I^2(2I-1)}\tau_2. \qquad (11.32)$$

This holds in the extreme narrowing approximation and under the assumption that the asymmetry parameter η is zero or negligible. Equation (11.32) is true for all molecular models.

In order to determine the coupling constant $e^2qQ/\hbar$ from (11.32) one needs the relaxation time T_1 and the correlation time τ_2 for spherical harmonics of rank 2. The first of these is determined experimentally but the second is obtained indirectly. To explain how this is done let us consider benzene C_6H_6 and perdeuterated benzene C_6D_6, which are circular plate molecules. The interactions with which we are now concerned are intramolecular. By employing the dilution technique of section 11.1 the spin-lattice rate of relaxation by dipolar interactions of a proton in C_6H_6 can be measured. Since ^{12}C has zero magnetic moment, the relaxation is due entirely to the interaction of the proton with the other five protons in C_6H_6. On account of the relative weakness of nuclear magnetic moments the five relaxation effects are additive.

We label by 1 the proton under consideration and by r_{1i} the distance from it to the ith proton in the molecule. On putting $I = \frac{1}{2}$ we obtain from (6.40) and (6.41) the rate of dipolar relaxation

$$\tfrac{3}{2}\gamma^4\hbar^2\tau_2 \sum_{i=2}^{6} r_{1i}^{-6}.$$

Equating this to the measured spin–lattice relaxation rate Bonera & Rigamonti (1965a) found that $\tau_2 = 2.4 \times 10^{-12}$ s for benzene at 22°C. Since the deuterons in C_6D_6 are in the same positions as the protons in C_6H_6, we can accept the same value of the orientational correlation time τ_2 for C_6D_6 and thus obtain (Bonera & Rigamonti, 1965b)

$$\frac{e^2qQ}{\hbar} = \pm 0.924\,\text{MHz}, \qquad q = \frac{1.45 \times 10^3}{4\pi\varepsilon_0}\,\text{nm}^{-3}.$$

Bonera & Rigamonti confirmed this by a theoretical study of the electric field gradient tensor which gave

$$q = \frac{1.6 \times 10^3}{4\pi\varepsilon_0} \, \text{nm}^{-3}, \qquad \eta = 0.007.$$

This small value of η justified the use of (11.32) in the above investigations.

11.7 Internal rotations

The theory developed in preceding chapters was based on the assumption that the rotating molecule, which contains the nucleus in whose relaxation we are interested, may be treated as a rigid body. This assumption produced values of relaxation times which could be compared with experimental findings and so lead to the values of quantities like correlation times, diffusion coefficients and the elements of interaction tensors.

However, comparatively few molecules in a liquid are rigid and over the past 25 years attempts have been made to establish a theory of internal rotations. Thus Woessner (1962*a,b*) calculated the correlation function for spherical harmonics of the second rank associated with the direction of a vector that rotates at a fixed angle to a principal coordinate axis of a rotating ellipsoid. Then Wallach (1967) examined the effect on orientational correlation functions of internal rotations in proteins.

We shall now give a brief account of some more recent theoretical investigations of internal rotations and their relation to available experimental results. We begin with the molecular coordinate frames S' and S'' of sections 8.1 and 9.3, which are fixed with respect to the molecule as a whole. In the cases of anisotropic chemical shift and quadrupole interaction we shall assume that the z''-axis is an axis of cylindrical symmetry for the relevant interaction tensor, so that the asymmetry parameter vanishes. We now take a third molecular coordinate frame S''', whose z'''-axis is the axis of the internal rotation that interests us. Calculations by Versmold (1970) show that the spin-lattice relaxation time in rotational diffusion theory and extreme narrowing approximation for a symmetric rotator molecule may be expressed as (Zeidler, 1974)

$$\frac{1}{T_1} = C \sum_{m,n=-2}^{2} \frac{d_{nm}^2(\beta')^2 d_{n0}^2(\theta'')^2}{6D_1 + m^2(D_3 - D_1) + n^2 D_{\text{int}}}, \tag{11.33}$$

where

$$C = \begin{cases} 2I(I+1)\gamma^4\hbar^2 r^{-6} & \text{for intramolecular dipolar interaction} \\ \frac{3}{10}\omega_0^2\delta_{z''}^2 & \text{for anisotropic chemical shift} \\ \dfrac{3}{40}\dfrac{2I+3}{I(2I-1)}\left(\dfrac{e^2qQ}{\hbar}\right)^2 & \text{for quadrupole interaction.} \end{cases}$$

In (11.33) D_1 and D_3 are given by (D.60), D_{int} is the diffusion coefficient for the z'''-axis, θ'' is the angle between the z''-axis and the z'''-axis, β' is the angle between the z'-axis and the z'''-axis, and the functions $d_{ll'}^2$ are defined in (B.19).

Apart from the rotational diffusive motion, finite steplike rotations through an angle $2\pi/3$ have been considered by Versmold and other researchers.

A considerable amount of theoretical and experimental investigation on internal rotations has been performed by Hertz and his collaborators. The model on which their calculations are based is that the non-rigid molecule consists of a collection of rigid molecules moving relative to one another. In rotational diffusion theory we found in (6.49) that under certain conditions the spectral density of $Y_{2m}(\theta, \phi)$ for a symmetric rotator is proportional to $2\tau/(1+\omega^2\tau^2)$. This is the Fourier transform of $\exp(-|t|/\tau)$, so this exponential is the normalized correlation function of $Y_{2m}(\theta, \phi)$ by the Wiener–Khinchin theorem (McConnell, 1980b, p. 71). It was therefore conjectured (Blicharska, Frech & Hertz, 1984) that the correlation function $g(t)$ of the spherical harmonic for the non-rigid molecule is expressible by

$$g(t) \approx \sum_j w_j \, e^{-t/\tau_j} \qquad (11.34)$$

summed over the constituent molecules subject to the condition

$$\sum_j w_j = 1, \qquad (11.35)$$

where the weight factors w_j are determined by the geometry of the non-rigid molecule. However, experiments on dimethylsulphoxide mixed with water produced results that conflicted with (11.35). In order to remedy this, (11.34) was replaced by a Cole–Davidson distribution of correlation times (Frech & Hertz, 1984).

The general picture is that at present there is no satisfactory quantitative theory of internal rotations of molecules in liquids (Frech & Hertz, 1985).

Appendix A

Representation of operators

In this appendix we collect certain well known results in operator theory and matrix algebra that are required for the main text and later appendixes without providing complete proofs for all the theorems mentioned (Born, 1957, Appendix XXV; McConnell, 1960, sections 13, 15, 16, 21, 23, 24; Schiff, 1968, Chap. 6).

We begin with a study of operators. Examples of these are the rotation operator related to the transformation of one set of cartesian coordinate axes to another with the same origin, and the Laplace operator ∇^2. The simplest example is the *identity operator* E which leaves any quantity f unaltered, so that $Ef = f$. The next simplest example is the operator aE, where a is a real or complex number, so that $aEf = af$ for all values of f. It may happen that an operator A has the property that for some values of f

$$Af = af. \tag{A.1}$$

When this is so, we say that f is an *eigenfunction* of A and that a is the corresponding *eigenvalue*. If A is a function of one or more real variables q, the *adjoint operator* A^+ of A is defined by

$$\int f_1^* A f_2 \, dq = \int (A^+ f_1)^* f_2 \, dq \tag{A.2}$$

for arbitrary functions f_1, f_2 of q, where in (A.2) the asterisk signifies complex conjugate and dq denotes the volume element in q-space. It is easily deduced from (A.2) that $A^{++} = A$ and that for any two operators A and B

$$(AB)^+ = B^+ A^+. \tag{A.3}$$

When $A^+ = A$, A is a *self-adjoint operator*. Examples of this are the cartesian coordinates x, y, z and the corresponding quantum mechanical linear momentum operators given by

$$p_x = -i\hbar \frac{\partial}{\partial x}, \qquad p_y = -i\hbar \frac{\partial}{\partial y}, \qquad p_z = -i\hbar \frac{\partial}{\partial z}.$$

If a self-adjoint operator A has an eigenfunction f with eigenvalue a, then by (A.1) and (A.2)

$$a \int f^* f \, dq = \int f^* A f \, dq = \int (A^+ f)^* f \, dq = a^* \int f^* f \, dq,$$

which shows that the eigenvalues of a self-adjoint operator are real. If f_1 and f_2 are two eigenfunctions of the self-adjoint A with distinct respective real eigenvalues k_1, k_2,

$$(k_1 - k_2) \int f_1^* f_2 \, dq = \int k_1^* f_1^* f_2 \, dq - \int f_1^* k_2 f_2 \, dq$$

$$= \int (A f_1)^* f_2 \, dq - \int f_1^* A f_2 \, dq = 0,$$

so that $\int f_1^* f_2 \, dq$ vanishes. By suitably multiplying the eigenfunctions $f_1, f_2, f_3, \ldots$ by numerical constants we can make $\int f_i^* f_i \, dq$ equal to unity and so have

$$\int f_i^* f_j \, dq = \delta_{ij}, \tag{A.4}$$

the Kronecker delta. The integral may be regarded as the scalar product of f_i and f_j, so that we can write

$$\int f_i^* f_j \, dq = (f_i, f_j). \tag{A.5}$$

When a set of functions satisfies (A.4), we say that they constitute an *orthonormal set*. The theory may be extended to cover discrete variables, like spin variables, and then the condition for orthonormality (A.4) is more properly expressed as

$$(f_i, f_j) = \delta_{ij}. \tag{A.6}$$

We shall be concerned only with *linear operators*; that is, those that satisfy

$$A(\phi + \psi) = A\phi + A\psi$$

for two arbitrary functions of q. When any arbitrary function of q is expressible as a linear combination of the members of an orthonormal set, we say that we have a *complete set*. If we write

$$\phi(q) = a_1 f_1 + a_2 f_2 + a_3 f_3 + \ldots, \tag{A.7}$$

where the a_i's are independent of q, and assume that we may integrate term by term, we deduce from (A.4) that

$$a_i = \int f_i^* \phi(q) \, dq. \tag{A.8}$$

For the case of $\phi(q) \equiv A f_i$ we write

$$A f_i = \sum_j A_{ji} f_j \tag{A.9}$$

and then, from (A.8),

$$A_{ji} = \int f_j^* A f_i \, dq. \tag{A.10}$$

This may also be written

$$A_{ji} = (f_j, A f_i) \tag{A.11}$$

in conformity with (A.5), or

$$A_{ji} = (f_j | A | f_i). \tag{A.12}$$

One easily deduces from (A.2) and (A.10) that

$$(A^+)_{ji} = (A_{ij})^* \tag{A.13}$$

and for two linear operators A and B that

$$(A + B)_{ji} = A_{ji} + B_{ji}$$

$$(AB)_{ji} = \sum_m A_{jm} B_{mi}.$$

Hence, the array of numbers defined by (A.10)–(A.12) obeys the rules of matrix addition and multiplication. We say that a complete orthonormal set constitutes the *basis* of a *representation*, and that (A.10) defines the *matrix representation* of the operator A with respect to this basis. The matrix with elements A_{ji} is the *matrix representative* of A. We see from (A.13) that the matrix representative of A^+ is the Hermitian conjugate of the matrix representative of A. Thus a self-adjoint operator is represented by a Hermitian matrix. Conversely, if an operator is represented by a Hermitian matrix for any basis, the operator is self-adjoint. It is obvious that the matrix representative of the identity operator is the *unit matrix* having ones along the diagonal and zeros elsewhere. We may regard the single column matrix with elements $a_1, a_2, a_3, \ldots$ as the representative of the function $\phi(q)$ in (A.7).

A *unitary operator* U is defined by the equations

$$U^+ U = U U^+ = E \tag{A.14}$$

and a *unitary matrix* with elements U_{ik} is defined by

$$\sum_l U_{il}^+ U_{lk} = \sum_l U_{il} U_{lk}^+ = \delta_{ik}. \tag{A.15}$$

If a basis $f_1, f_2, \ldots$ is transformed to the basis $f'_1, f'_2, \ldots$ by

$$f'_i = \sum_j U_{ji} f_j, \tag{A.16}$$

it may readily be deduced from (A.4) applied also to $f'_1, f'_2, \ldots$ and from (A.16) that U_{ji} satisfies (A.15) and so the corresponding matrix is unitary. A *unitary transformation* of an operator B is defined by

$$B' = U^+ B U, \tag{A.17}$$

where U satisfies (A.14). The *trace* of any operator is the sum of the diagonal elements of its matrix representative. If A and B are two operators and we define the matrix elements A_{ik}, B_{ik} with reference to a finite basis $f_1, f_2, \ldots, f_n$, then

$$\mathrm{tr}(AB) = \sum_{k=1}^{n} (AB)_{kk} = \sum_{k,l=1}^{n} A_{kl} B_{lk} = \sum_{k,l=1}^{n} B_{lk} A_{kl} = \mathrm{tr}(BA). \tag{A.18}$$

Hence, $\mathrm{tr}(AB - BA)$ vanishes in any finite-dimensional representation. From (A.14), (A.17) and (A.18)

$$\mathrm{tr}\, B' = \mathrm{tr}(U^+ B U) = \mathrm{tr}(U U^+ B) = \mathrm{tr}(EB) = \mathrm{tr}\, B, \tag{A.19}$$

which shows that the trace is independent of the basis of the finite representation.

Suppose that we have an algebraic relation between operators $B_1, B_2, B_3, \ldots$; for example,

$$\lambda_1 B_1^2 B_2 + \lambda_2 B_3 B_4 B_5 = B_1 B_3 B_5. \tag{A.20}$$

Then on transforming according to (A.17) we obtain

$$\lambda_1 B_1'^2 B_2' + \lambda_2 B_3' B_4' B_5' = B_1' B_3' B_5',$$

which shows that (A.20) is invariant under a unitary transformation. Applying this to a finite-dimensional representation we deduce that, if we have an algebraic relation between the matrix representatives of $B_1, B_2, B_3, \ldots$, this relation will persist, if the basis of the representation is altered.

Appendix B

The rotation operator

We shall define in this appendix spherical harmonics, Wigner functions, the rotation operator, Euler angles and spherical tensors, and we shall recall some theorems needed for applications to nuclear magnetic relaxation problems.

The *spherical harmonic* $Y_{lm}(\theta, \phi)$ is defined for integers l, m by

$$Y_{lm}(\theta, \phi) = \left[\frac{2l+1}{4\pi} \frac{(l-m)}{(l+m)!}\right]^{1/2} \frac{e^{im\phi}(-\sin \theta)^m}{2^l l!} \left[\frac{d}{d(\cos \theta)}\right]^{l+m} (\cos^2 \theta - 1)^l.$$

(B.1)

From this definition it appears that Y_{lm} vanishes unless $|m| \leqslant l$, and therefore there exist $2l + 1$ spherical harmonics

$$Y_{l,-l}(\theta, \phi), Y_{l,-l+1}(\theta, \phi), Y_{l,-l+2}(\theta, \phi), \ldots, Y_{l,l-1}(\theta, \phi), Y_{ll}(\theta, \phi). \text{(B.2)}$$

Among the more important properties of spherical harmonics we may note (Edmonds, 1968; Rose, 1957)

$$Y_{lm}^*(\theta, \phi) = (-)^m Y_{l,-m}(\theta, \phi)$$ (B.3)

$$Y_{lm}(-\theta, -\phi) = (-)^m Y_{lm}^*(\theta, \phi)$$ (B.4)

$$\int_0^{2\pi} d\phi \int_0^\pi Y_{lm}^*(\theta, \phi) Y_{l'm'}(\theta, \phi) \sin \theta \, d\theta = \delta_{ll'} \delta_{mm'}.$$ (B.5)

The spherical harmonics play a large part in the study of rotation of a rigid body in three-dimensional space. To discuss this we introduce the rotation operator.

Let us take a set of rectangular coordinate axes Ox, Oy, Oz fixed with respect to the laboratory and a function $f(x, y, z)$ of the coordinates (x, y, z) of a point. We rotate the coordinate axes about O denoting the coordinates of the same point by (x', y', z') and we form the same function f of the new coordinates. Expressing the relation of the function of the two triads of coordinates by

$$f(x', y', z') = Rf(x, y, z)$$ (B.6)

we say that R is the *rotation operator* for the prescribed rotation of axes and for the function f. Usually f will have more than one component and then (B.6) is to be interpreted as an abbreviation for

$$f_i(x', y', z') = \sum_k R_{ki} f_k(x, y, z).$$ (B.7)

The *infinitesimal generators of rotation* J_x, J_y, J_z are the operators defined by

$$J_x = -i\left(y\frac{\partial}{\partial z} - z\frac{\partial}{\partial y} \right), \ J_y = -i\left(z\frac{\partial}{\partial x} - x\frac{\partial}{\partial z} \right), \ J_z = -i\left(x\frac{\partial}{\partial y} - y\frac{\partial}{\partial x} \right).$$ (B.8)

Since

$$J_x = \hbar^{-1}(yp_z - zp_y),$$ (B.9)

etc., where p_y, p_z are self-adjoint components of quantum mechanical linear momentum of a particle, it follows from (A.3) that J_x, and similarly J_y and J_z, are self-adjoint operators. In terms of spherical polar coordinates r, θ, ϕ given by

$$x = r\sin\theta\cos\phi, \qquad y = r\sin\theta\sin\phi, \qquad z = r\cos\theta$$ (B.10)

we have

$$J_z = -i\frac{\partial}{\partial\phi}.$$

Then writing $f(x, y, z)$ as $g(r, \theta, \phi)$ we find for a rotation about the z-axis through an infinitesimal angle ε that

$$Rg(r, \theta, \phi) = g(r, \theta, \phi - \varepsilon) = (E - i\varepsilon J_z)g(r, \theta, \phi).$$

By iteration one immediately deduces that the operator for a rotation through a finite angle χ about the z-axis is $\exp(-i\chi J_z)$ and about an axis specified by the unit vector $\mathbf{e}$ is R given by

$$R = \exp[-i\chi(\mathbf{J}\cdot\mathbf{e})].$$ (B.11)

The angle χ for a rotating body will be a function of the time t. Expanding the exponential and differentiating term by term we deduce from (B.11) that

$$\frac{dR(t)}{dt} = -i(\mathbf{J}\cdot\boldsymbol{\omega}(t))R(t),$$ (B.12)

where $\boldsymbol{\omega}(t)$ is the instantaneous angular velocity. For thermal motion, χ and ω will be random variables. The same will be true for $R(t)$, which is then called a *stochastic rotation operator*.

The most general rotation of a rigid body may be obtained by taking the origin of the laboratory frame of axes through the point of rotation, rotating the body about the z-axis through an angle γ, then about the y-axis through an angle β and again about the z-axis through an angle α (Rose, 1957). These are the *Euler angles* shown in Fig. B.1. From (B.11) the rotation operator is given by

$$R = \exp(-i\alpha J_z)\exp(-i\beta J_y)\exp(-i\gamma J_z). \tag{B.13}$$

Since the infinitesimal generators J_x, J_y, J_z are self-adjoint, (A.3) gives

$$R^+ = \exp(i\gamma J_z)\exp(i\beta J_y)\exp(i\alpha J_z). \tag{B.14}$$

Hence,

$$RR^+ = R^+R = E, \tag{B.15}$$

so that, by (A.14), R is a unitary operator.

Let us link the rotation operator of (B.13) with the spherical harmonics of (B.2). In analogy with (B.7) we write

$$Y_{lm}(\theta', \phi') = \sum_{m'=-l}^{l} R^l_{m'm} Y_{lm'}(\theta, \phi). \tag{B.16}$$

We see, from (B.5) with $l' = l$ and the fact that $\sin\theta\, d\theta\, d\phi$ is the volume element, that the $Y_{ls}(\theta, \phi)$ constitute an orthonormal set which we shall

Fig. B.1 Orientation of a rigid body as a result of successive rotations through angle γ about Oz, angle β about Oy and angle α about Oz.

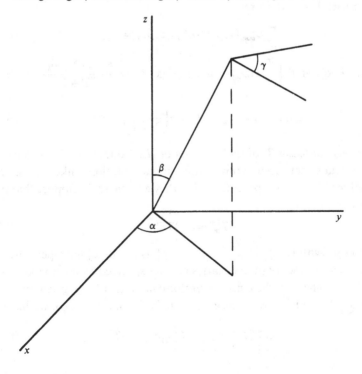

assume to be complete. Then, from (A.9) and (A.10), $R^l_{m'm}$ is the $m'm$-element of the R of (B.13) in the representation with basis (B.2). In the notation of Rose (1957), (B.16) is written

$$Y_{lm}(\theta', \phi') = \sum_{m'=-l}^{l} D^l_{m'm}(\alpha, \beta, \gamma) Y_{lm'}(\theta, \phi), \tag{B.17}$$

where $D^l_{m'm}(\alpha, \beta, \gamma)$ is the *Wigner function* defined by

$$D^l_{m'm}(\alpha, \beta, \gamma) = e^{-i(m'\alpha + m\gamma)} d^l_{m'm}(\beta), \tag{B.18}$$

$$d^l_{m'm}(\beta) = [(l+m)! \, (l-m')! \, (l+m')! \, (l-m)!]^{1/2}$$
$$\times \sum_{s} \frac{(-)^s (\cos \tfrac{1}{2}\beta)^{2l+m-m'-2s} (-\sin \tfrac{1}{2}\beta)^{m'-m+2s}}{(l-m'-s)! \, (l+m-s)! \, (s+m'-m)! \, s!} \tag{B.19}$$

and the sum is over all values of s for which the arguments of the factorials are non-negative.

If we compare (B.13) and (B.14), we see that we may go from R to R^+ by making the substitutions

$$\alpha \mapsto -\gamma, \qquad \beta \mapsto -\beta, \qquad \gamma \mapsto -\alpha.$$

On employing (A.13) we deduce that

$$D^l_{mm'}(-\gamma, -\beta, -\alpha) = D^{l*}_{m'm}(\alpha, \beta, \gamma). \tag{B.20}$$

Other useful relations are

$$\sum_{m'} D^l_{mm'}(\alpha, \beta, \gamma) D^{l*}_{m''m'}(\alpha, \beta, \gamma) = \delta_{mm''} \tag{B.21}$$

$$\int_0^{2\pi} d\gamma \int_0^{\pi} d\beta \sin \beta \int_0^{2\pi} d\alpha \, D^{l*}_{mk}(\alpha, \beta, \gamma) D^{l'}_{m'k'}(\alpha, \beta, \gamma) = \frac{8\pi^2 \delta_{ll'} \delta_{mm'} \delta_{kk'}}{2l+1} \tag{B.22}$$

$$D^l_{0m}(\alpha, \beta, \gamma) = D^l_{0m}(0, \beta, \gamma) = \left(\frac{4\pi}{2l+1}\right)^{1/2} Y^*_{lm}(\beta, \gamma). \tag{B.23}$$

A *spherical tensor* $\mathbf{T}^l$ of *rank* l is a set of $2l+1$ quantities T^l_m with $|m| \leqslant l$, which transform under three-dimensional rotations like (B.2). The quantities T^l_m may be operators and, if they are, we shall suppose that they obey the relation

$$T^{l+}_m = (-)^m T^l_{-m} \tag{B.24}$$

in analogy with (B.3). This shows that T^l_0 is a self-adjoint operator.

Let us make the rotation of cartesian axes specified by the operator R of (B.13). The effect on the tensor operator component T^l_m is given by (A.17) with $B \equiv T^l_m$ and $U \equiv R^+$. Since T^l_m transforms like $Y_{lm}(\theta, \phi)$, we have

$$R T^l_m R^+ = \sum_{m'=-l}^{l} D^l_{m'm}(\alpha, \beta, \gamma) T^l_{m'}. \tag{B.25}$$

This equation defines an *irreducible tensor operator* of rank l. We see from (B.17) that

$$Y^*_{lm}(\theta', \phi') = \sum_{m'=-l}^{l} D^{l*}_{m'm}(\alpha, \beta, \gamma) Y^*_{lm'}(\theta, \phi),$$

and so for a spherical tensor operator component V^l_m

$$R V^{l+}_m R^+ = \sum_{n'=-l}^{l} D^{l*}_{n'm}(\alpha, \beta, \gamma) V^{l+}_{n'}. \tag{B.26}$$

Hence, from (B.15),

$$R \sum_m V^{l+}_m T^l_m R^+ = \sum_m R V^{l+}_m R^+ R T^l_m R^+$$

$$= \sum_{m,n'} D^{l*}_{n'm}(\alpha, \beta, \gamma) D^l_{m'm}(\alpha, \beta, \gamma) V^{l+}_{n'} T^l_{m'} = \sum_{m'} V^{l+}_{m'} T^l_{m'},$$

by (B.21). Thus $\sum_{m=-l}^{l} V^l_m T^l_m$ is invariant under three-dimensional rotations. We therefore define it as the *scalar product* of $\mathbf{V}^l$ and $\mathbf{T}^l$. Since these are operators, their order must be preserved. Denoting the scalar product by $(\mathbf{V}^l \cdot \mathbf{T}^l)$ and employing (B.24) we write

$$(\mathbf{V}^l \cdot \mathbf{T}^l) = \sum_{q=-l}^{l} (-)^q V^l_{-q} T^l_q. \tag{B.27}$$

We may argue conversely that, if $\sum_{q=-l}^{l} V^l_q T^l_q$ is invariant under three-dimensional rotations and if $\mathbf{T}^l$ is an irreducible spherical tensor operator, it will follow that $\mathbf{V}^l$ is also an irreducible spherical tensor operator. Indeed we have by supposition and (B.25)

$$\sum_q V^{l+}_q T^l_q = R \sum_s V^{l+}_s T^l_s R^+ = \sum_s R V^{l+}_s R^+ R T^l_s R^+$$

$$= \sum_s R V^{l+}_s R^+ \sum_q D^l_{qs}(\alpha, \beta, \gamma) T^l_q.$$

Assuming that the T^l_m form a complete set, we may identify the coefficients of T^l_q on both sides and therefore

$$V^{l+}_q = \sum_s R V^{l+}_s R^+ D^l_{qs}(\alpha, \beta, \gamma)$$

$$R^+ V^{l+}_q R = \sum_s V^{l+}_s D^l_{qs}(\alpha, \beta, \gamma).$$

We see from (B.20) that, when we interchange R and R^+ on the left hand side, we replace $D^l_{qs}(\alpha, \beta, \gamma)$ by $D^{l*}_{sq}(\alpha, \beta, \gamma)$ so that

$$R V^{l+}_q R^+ = \sum_s V^{l+}_s D^{l*}_{sq}(\alpha, \beta, \gamma).$$

On taking the adjoint of both sides we deduce that

$$R V^l_q R^+ = \sum_s D^l_{sq}(\alpha, \beta, \gamma) V^l_s.$$

and this, by (B.25), establishes that V^l is an irreducible spherical tensor operator of rank l. This result will be very useful for the discussion of interaction Hamiltonians that describe nuclear magnetic relaxation processes.

We shall give explicit values of the spherical harmonics for low values of l. Putting $l=0$, so that $m=0$, in (B.1) we obtain

$$Y_{00}(\theta, \phi) = (4\pi)^{-1/2}$$

which trivially satisfies (B.5). For $l=1$ we find that

$$Y_{1,-1}(\theta, \phi) = \left(\frac{3}{8\pi}\right)^{1/2} \sin \theta \, e^{-i\phi}$$

$$Y_{10}(\theta, \phi) = \left(\frac{3}{4\pi}\right)^{1/2} \cos \theta \tag{B.28}$$

$$Y_{11}(\theta, \phi) = -\left(\frac{3}{8\pi}\right)^{1/2} \sin \theta \, e^{i\phi},$$

which clearly satisfy (B.4) and (B.5). We see from (B.10) and (B.28) that r_{-1}, r_0, r_1 defined by

$$r_{-1} = \frac{x-iy}{2^{1/2}}, \qquad r_0 = z, \qquad r_1 = -\frac{x+iy}{2^{1/2}}$$

are components of a spherical tensor of rank one. The same is true for A_{-1}, A_0, A_1 defined by

$$A_{-1} = \frac{A_x - iA_y}{2^{1/2}}, \qquad A_0 = A_z, \qquad A_1 = -\frac{A_x + iA_y}{2^{1/2}}, \tag{B.29}$$

where A_x, A_y, A_z are the cartesian components of a vector. These components may be operators; for example, they could be the J_x, J_y, J_z defined by (B.8).

Putting $l=2$ in (B.1) we have

$$Y_{2,\pm 1}(\theta, \phi) = \mp\left(\frac{15}{8\pi}\right)^{1/2} \cos \theta \sin \theta \, e^{\pm i\phi}$$

$$Y_{20}(\theta, \phi) = \left(\frac{5}{16\pi}\right)^{1/2} (3\cos^2 \theta - 1) \tag{B.30}$$

$$Y_{2,\pm 2}(\theta, \phi) = \left(\frac{15}{32\pi}\right)^{1/2} \sin^2 \theta \, e^{\pm 2i\phi}.$$

From these we deduce that $r_{2,-2}$, $r_{2,-1}$, r_{20}, r_{21}, r_{22} given by

$$r_{2,\pm 1} = \mp\left(\frac{15}{8\pi}\right)^{1/2} z(x \pm iy), \qquad r_{20} = \left(\frac{5}{16\pi}\right)^{1/2} (3z^2 - r^2)$$

$$r_{2,\pm 2} = \left(\frac{15}{32\pi}\right)^{1/2} (x \pm iy)^2$$

constitute a spherical tensor of rank 2.

Appendix C

Construction of spherical tensors

We shall first apply the quantum theory of angular momentum to the infinitesimal generators of rotation. Then we shall introduce Clebsch–Gordan coefficients and employ them to construct tensor components for irreducible representations of the three-dimensional rotation group. Finally we shall discuss the Wigner–Eckart theorem and its application to the calculation of correlation functions.

We see from (B.8) and (B.9) that the quantum mechanical orbital angular momentum $\mathbf{M}$ of a particle is related to $\mathbf{J}$ by

$$\mathbf{M} = \hbar \mathbf{J}. \tag{C.1}$$

Corresponding to the commutation relations for the components of $\mathbf{M}$ we have

$$[J_y, J_z] = \mathrm{i} J_x, \qquad [J_z, J_x] = \mathrm{i} J_y, \qquad [J_x, J_y] = \mathrm{i} J_z. \tag{C.2}$$

We shall now extend the definition of infinitesimal generator of the rotation group beyond (B.8) and say that it is any vector $\mathbf{J}$, whose components are self-adjoint operators satisfying (C.2). If we work in a representation of rank j, and therefore in a $2j+1$-dimensional representation, the eigenvalues of J_z are $-j, -j+1, \ldots, j-1, j$ and the eigenvalue of J^2 is $j(j+1)$. Moreover it is possible to obtain simultaneous eigenfunctions of J^2 and J_z.

Let us take two independent operators $\mathbf{J}_1$ and $\mathbf{J}_2$ and let us write

$$\mathbf{J} = \mathbf{J}_1 + \mathbf{J}_2. \tag{C.3}$$

Then each component of $\mathbf{J}_1$ commutes with each component of $\mathbf{J}_2$ and it is immediately deduced that $\mathbf{J}$ satisfies (C.2). We write $|j_1 m_1\rangle$ for a normalized simultaneous eigenfunction of J_1^2 and J_{1z} and $|j_2 m_2\rangle$ for a normalized simultaneous eigenfunction of J_2^2 and J_{2z}, where $-j_1 \leqslant m_1 \leqslant j_1$, $-j_2 \leqslant m_2 \leqslant j_2$. The possible values of j resulting from the addition of $\mathbf{J}_1$ and $\mathbf{J}_2$ in (C.3) are

$$j_1 + j_2, j_1 + j_2 - 1, \ldots, |j_1 - j_2| \tag{C.4}$$

and each of these provides an irreducible representation (Condon &

Shortley, 1935, Chap. III; McConnell, 1960, section 24). The geometrical interpretation of (C.4) is that lines of lengths j_1, j_2 and j can form a triangle. This condition, which is clearly symmetric in j_1, j_2, j, is denoted by $\Delta(j_1 j_2 j)$.

We now write

$$|j_1 j_2 m_1 m_2) \equiv |j_1 m_1\rangle|j_2 m_2).$$

Suppose that we combine $\mathbf{J}_1$ and $\mathbf{J}_2$ so that j has one of the values (C.4). We denote by $|j_1 j_2 jm)$ the normalized wave function of the state specified by the eigenvalues of J_1^2, J_2^2, J^2 and J_z, and we expand it as a series of the $|j_1 j_2 m_1 m_2)$:

$$|j_1 j_2 jm) = \sum (j_1 j_2 m_1 m_2 | j_1 j_2 jm)|j_1 j_2 m_1 m_2) \qquad (C.5)$$

summed over the states such that m is $m_1 + m_2$ and $\Delta(j_1 j_2 j)$. In the notation of (A.7), (A.8) and (A.11)

$$\phi(q) \mapsto |j_1 j_2 jm), \qquad f_i \mapsto |j_1 j_2 m_1 m_2)$$
$$(f_i, \phi(q)) \mapsto (j_1 j_2 m_1 m_2 | j_1 j_2 jm).$$

It may be shown that (Racah, 1962)

$$(j_1 j_2 m_1 m_2 | j_1 j_2 jm) = \delta_{m, m_1 + m_2} \left[\frac{(2j+1)(j_1 + j_2 - j)!(j + j_1 - j_2)!(j + j_2 - j_1)!}{(j + j_1 + j_2 + 1)!} \right]^{1/2}$$
$$\times \sum_v \frac{(-)^v}{v!} \frac{[(j_1 + m_1)!(j_1 - m_1)!(j_2 + m_2)!(j_2 - m_2)!(j + m)!(j - m)!]^{1/2}}{(j_1 + j_2 - j - v)!(j_1 - m_1 - v)!(j_2 + m_2 - v)!(j - j_2 + m_1 + v)!(j - j_1 - m_2 + v)!}, (C.6)$$

the summation over v being over the integers such that the arguments of the factorials are non-negative. The coefficient of $|j_1 j_2 m_1 m_2)$ in (C.5) is called a *Clebsch–Gordan coefficient*, or C–G coefficient. It is also written more compactly as $C(j_1 j_2 j; m_1 m_2 m)$.

Let us take two commuting tensor operators $\mathbf{U}^{j_1}$ of rank j_1 and $\mathbf{T}^{j_2}$ of rank j_2. Then

$$V_m^j = \sum C(j_1 j_2 j; m_1 m_2 m) U_{m_1}^{j_1} T_{m_2}^{j_2} \qquad (C.7)$$

summed over the allowed values of m_1, m_2, j_1, j_2 defines $\mathbf{V}^j$ as the *tensor product* of $\mathbf{U}^{j_1}$ and $\mathbf{T}^{j_2}$. For each value of j in (C.4) there will be an irreducible representation of the rotation group and the number of elements of $\mathbf{V}^j$ will be $2j + 1$. In future applications we shall be concerned with the cases of $j_1 = j_2 = 1$, so that j may have the values 2, 1, 0.

As an illustration let us calculate V_1^2 from (C.7). We then have

$$j_1 = j_2 = 1; \qquad j = 2, m = 1; \qquad m_1 = 0, m_2 = 1 \text{ or } m_1 = 1, m_2 = 0.$$

We see from (C.6) that

$$C(112; 011) = \left[\frac{5 \cdot 0! \, 2! \, 2!}{5!} \right]^{1/2} \sum_v \frac{(-)^v}{v!} \frac{[1! \, 1! \, 2! \, 0! \, 3! \, 1!]^{1/2}}{(-v)!(1-v)!(2-v)!(1+v)! \, v!}.$$

The only allowable value of v is zero, and therefore
$$C(112;011)=2^{-1/2}.$$
Similarly $C(112; 101)$ is equal to $2^{-1/2}$, and therefore (C.7) yields
$$V_1^2=2^{-1/2}(U_0^1T_1^1+U_1^1T_0^1). \tag{C.8}$$
We likewise find that
$$V_{-1}^2=2^{-1/2}(U_0^1T_{-1}^1+U_{-1}^1T_0^1). \tag{C.9}$$
Proceeding in the same way we obtain
$$V_2^2=U_1^1T_1^1, \qquad V_{-2}^2=U_{-1}^1T_{-1}^1 \tag{C.10}$$
$$V_0^2=6^{-1/2}(U_1^1T_{-1}^1+U_{-1}^1T_1^1)+(\tfrac{2}{3})^{1/2}U_0^1T_0^1. \tag{C.11}$$
On writing, as in (B.29),
$$U_{-1}^1=\frac{U_x^1-iU_y^1}{2^{1/2}}, \qquad U_0^1=U_z^1, \qquad U_1^1=-\frac{U_x^1+iU_y^1}{2^{1/2}}$$
$$T_{-1}^1=\frac{T_x^1-iT_y^1}{2^{1/2}}, \qquad T_0^1=T_z^1, \qquad T_1^1=-\frac{T_x^1+iT_y^1}{2^{1/2}}$$
we express (C.11) as
$$V_0^2=6^{-1/2}\{3U_z^1T_z^1-(\mathbf{U}^1\cdot\mathbf{T}^1)\}. \tag{C.12}$$
When $j=1$, we find that
$$V_1^1=2^{-1/2}(U_1^1T_0^1-U_0^1T_1^1) \tag{C.13}$$
$$V_{-1}^1=2^{-1/2}(U_0^1T_{-1}^1-U_{-1}^1T_0^1) \tag{C.14}$$
$$V_0^1=2^{-1/2}(U_1^1T_{-1}^1-U_{-1}^1T_1^1). \tag{C.15}$$
When $j=0$, we have a tensor with a single component
$$V_0^0=-3^{-1/2}\{U_0^1T_0^1-U_1^1T_{-1}^1-U_{-1}^1T_1^1\}. \tag{C.16}$$
This may be written in the alternative forms:
$$V_0^0=-3^{-1/2}\sum_{q=-1}^{1}(-)^qU_{-q}^1T_q^1 \tag{C.17}$$
$$V_0^0=-3^{-1/2}(\mathbf{U}^1\cdot\mathbf{T}^1). \tag{C.18}$$
For the special case of $\mathbf{U}^1\equiv\mathbf{T}^1\equiv\mathbf{I}$ (C.8)–(C.12) become
$$V_{\pm1}^2=2^{-1/2}(I_0I_{\pm1}+I_{\pm1}I_0) \tag{C.19}$$
$$V_{\pm2}^2=I_{\pm1}^2 \tag{C.20}$$
$$V_0^2=6^{-1/2}(I_1I_{-1}+I_{-1}I_1)+(\tfrac{2}{3})^{1/2}I_0^2$$
$$=6^{-1/2}\{3I_z^2-I(I+1)\}. \tag{C.21}$$

On writing

$$I_+ = I_x + iI_y, \qquad I_- = I_x - iI_y, \qquad I_0 = I_z, \tag{C.22}$$

so that by (B.29)

$$I_+ = -2^{1/2}I_1, \qquad I_- = 2^{1/2}I_{-1},$$

(C.19)–(C.21) become

$$V^2_{\pm 1} = \mp\tfrac{1}{2}(I_z I_\pm + I_\pm I_z) \tag{C.23}$$

$$V^2_{\pm 2} = \tfrac{1}{2}I^2_\pm \tag{C.24}$$

$$V^2_0 = 6^{-1/2}\{3I^2_z - I(I+1)\}. \tag{C.25}$$

By taking the Euler angles to be infinitesimal it may be verified from (C.23)–(C.25) that V^2_m satisfies (B.25).

Another application of the C–G coefficients is to be found in the Wigner–Eckart theorem (Eckart, 1930; Rose, 1957; Wigner, 1959). Let us take a spherical tensor operator $\mathbf{T}^L$ of rank L and let T^L_M be its Mth component. The *Wigner–Eckart theorem* states that the matrix element

$$(j'm'|T^L_M|jm) = C(jLj'; mMm')(j'\|\mathbf{T}^L\|j), \tag{C.26}$$

where we have used the notation of (A.12) for the matrix element on the left hand side. The quantity $(j'\|\mathbf{T}^L\|j)$ is called the *reduced matrix element* of the set T^L_M, and its value is independent of m, m', M. In order that the C–G coefficient be non-vanishing we must have $\Delta(jLj')$ and $m' = m + M$. We see that the C–G coefficient is the same for all operators of rank L. It may therefore be calculated by taking any convenient operator in (C.26). If $\mathbf{T}^L$ involves other operators denoted by b which commute with each other, with J^2 and with J_z, (C.26) is generalized to

$$(b'j'm'|T^L_M|bjm) = C(jLj'; mMm')(b'j'\|\mathbf{T}^L\|bj). \tag{C.27}$$

As an illustration of the usefulness of the Wigner–Eckart theorem let us consider the ensemble average $\langle S_{l'm'}(0)T_{lm}(t)\rangle$ for two tensor operator components of ranks l' and l. Hubbard (1969) expressed the ensemble average as $\mathrm{tr}(\rho S_{l'm'}(0)T_{lm}(t))$, where ρ is the density operator, applied the Wigner–Eckart theorem and employing properties of the C–G coefficients obtained the result

$$\langle S_{l'm'}(0)T_{lm}(t)\rangle = \delta_{l'l}\delta_{m',-m}(-)^m\langle S_{l0}(0)T_{l0}(t)\rangle. \tag{C.28}$$

Using (B.24) we deduce from (C.28) that

$$\langle S^+_{l'm}(0)T_{lm}(t)\rangle = \delta_{ll'}\langle S_{l0}(0)T_{l0}(t)\rangle. \tag{C.29}$$

In most of our calculations S and T will be commuting quantities, not operators, so that the adjoint is replaced by the complex conjugate. Then

(C.29) becomes

$$\langle S_{l'm}^*(0)T_{lm}(t)\rangle = \delta_{ll'}\langle S_{l0}(0)T_{l0}(t)\rangle. \tag{C.30}$$

The results (C.28)–(C.30) are true for S_{lm}, T_{lm} which may involve not only orientational variables like angles but also other variables denoted in (C.27) by b. These might, for example, be angular velocity components. If we apply the results to rotational Brownian motion and take S_{lm} and T_{lm} to be random functions depending only on orientational coordinates, it may be possible to derive an expression for the ensemble average $\langle S_{l0}(0)T_{l0}(t)\rangle$. For definiteness we take a rotating molecule and choose a body frame of coordinate axes through the point of rotation. Let θ', ϕ' be the spherical polar angles in the body frame of a vector from the origin and in a direction fixed with respect to the body. We also take a laboratory coordinate system with origin at the centre of rotation. We denote by $\theta(t)$, $\phi(t)$ the spherical polar angles of the above vector at time t with respect to the laboratory system and by $\theta(0)$, $\phi(0)$ the corresponding angles for time zero. It has been shown (Ford *et al.*, 1979, Appendix) that the ensemble average

$$\langle Y_{lq}^*(\theta(0), \phi(0))Y_{lq'}(\theta(t), \phi(t))\rangle$$

$$= \frac{\delta_{qq'}}{2l+1} \sum_{n,n'=-l}^{l} Y_{ln}(\theta', \phi')\langle R^+(t)\rangle_{nn'} Y_{ln'}^*(\theta', \phi'). \tag{C.31}$$

In this, $R(t)$ is the operator for the rotation of the molecule from its orientation at time zero to its orientation at time t and $R^+(t)$ is its adjoint. The average $\langle R^+(t)\rangle$ is calculated subject to the condition that $R(0)$ is the identity operator. The suffixes nn' denote the matrix element in the representation having the spherical harmonics of rank l in the body frame as basis.

The result (C.31) may be extended to spherical tensors whose components are commuting functions of orientational variables only. We denote by $\mathbf{a}$ a unit vector specifying a direction fixed with respect to the body frame and by $\mathbf{a}(t)$ the unit vector at time t specifying the same direction with respect to a laboratory system of coordinates. Then the proof employed for the derivation of (C.31) also holds to establish for a spherical tensor with components $H_{lq}(\mathbf{a}(t))$ the relation

$$\langle H_{lq}^*(\mathbf{a}(0))H_{lq'}(\mathbf{a}(t))\rangle = \frac{\delta_{qq'}}{2l+1} \sum_{n,n'=-l}^{l} \langle R^+(t)\rangle_{nn'} H_{ln'}^*(\mathbf{a})H_{ln}(\mathbf{a}). \tag{C.32}$$

We see from (C.32) that the cross-correlation function defined in section 2.3 vanishes, that the autocorrelation function is independent of q and that, since H_{l0} is real by (B.24),

$$\langle H_{l0}(\mathbf{a}(0)) H_{l0}(\mathbf{a}(t)) \rangle = \frac{1}{2l+1} \sum_{n,n'=-l}^{l} \langle R^+(t) \rangle_{nn'} H_{ln'}^*(\mathbf{a}) H_{ln}(\mathbf{a}). \quad (\text{C.33})$$

For compactness we shall often express (C.32) and (C.33) as

$$\langle H_{lq}^*(0) H_{lq'}(t) \rangle = \frac{\delta_{qq'}}{2l+1} \sum_{n,n'=-l}^{l} \langle R^+(t) \rangle_{nn'} H_{ln'}'^* H_{ln}' \quad (\text{C.34})$$

$$\langle H_{l0}(0) H_{l0}(t) \rangle = \frac{1}{2l+1} \sum_{n,n'=-l}^{l} \langle R^+(t) \rangle_{nn'} H_{ln'}'^* H_{ln}'. \quad (\text{C.35})$$

Appendix D

Spectral densities for molecular models

This appendix will be devoted to describing in a general way how the theory of steady state rotational Brownian motion is applied to the derivation of analytical expressions for correlation functions and spectral densities, which will be required in order to provide rates of nuclear magnetic relaxation due to intramolecular interactions of various types. In addition we shall list results of previous Brownian motion calculations that will be useful for our present investigations.

The starting point is the Euler–Langevin equations (2.57) for a rotating molecule. These have been studied for different molecular models with the purpose of deducing the autocorrelation function for a component of angular velocity. To find orientational correlation functions for physical quantities related to the rotating molecule we need a joint probability density function for two different times, say, time zero and time t. As pointed out in sections 2.1 and 2.2, the joint probability density function may be expressed as the product of an initial probability density function and a conditional probability density function. If then we consider the rotation of a molecule about its centre of mass and describe its orientation with respect to a laboratory coordinate system by three Euler angles with values $\alpha(0), \beta(0), \gamma(0)$ at time zero and $\alpha(t), \beta(t), \gamma(t)$ at time t, we can treat the conditional probability by means of the rotation operator $R(t)$, which is equal to the identity at time zero and later varies in a random manner due to the thermal motion of the environment of the molecule. To perform the ensemble averaging required for the orientational correlation function we later average over $\alpha(0), \beta(0), \gamma(0)$.

We must therefore solve (B.12) for $R(t)$ with the initial condition $R(0) = E$. We assume that the solution at a later time t consists of a slowly varying average $\langle R(t) \rangle$ about which there are small random variations, and we write

$$R(t) = (E + \varepsilon F^{(1)}(t) + \varepsilon^2 F^{(2)}(t) + \varepsilon^3 F^{(3)}(t) + \cdots)\langle R(t) \rangle, \qquad \text{(D.1)}$$

where ε is a small dimensionless parameter and the $F^{(i)}(t)$ are random functions. When $R(t)$ appears in an expression enclosed by angular brackets, it will be understood that the average is calculated subject to the above initial condition. We also assume that the non-stochastic $\langle R(t) \rangle$ itself obeys an equation

$$d\langle R(t) \rangle/dt = (\varepsilon \Omega^{(1)}(t) + \varepsilon^2 \Omega^{(2)}(t) + \varepsilon^3 \Omega^{(3)}(t) + \cdots)\langle R(t) \rangle, \qquad (D.2)$$

where the $\Omega^{(i)}(t)$ are non-stochastic functions. On solving between (B.12), (D.1), (D.2) and using the values of the correlation functions of the angular velocity components deduced from (2.57) it may be possible to find the values of $\varepsilon F^{(1)}(t), \varepsilon^2 F^{(2)}(t), \ldots, \varepsilon \Omega^{(1)}(t), \varepsilon^2 \Omega^{(2)}(t), \ldots$ and to solve (D.2) for $\langle R(t) \rangle$. This is all that is required for (C.31)–(C.35). If we need $R(t)$, as in fact we do for spin-rotational interactions, we obtain it from (D.1) and the known value of $\langle R(t) \rangle$.

It will be found that the Hamiltonian $\hbar G(t)$ for the interactions that give rise to nuclear magnetic relaxation is nearly always expressible by

$$\hbar G(t) = \hbar \sum_{q=-l}^{l} (-)^q H_{-q}(t) A_q, \qquad (D.3)$$

where A_q is the qth component of a spherical tensor operator of rank l, and $H_q(t)$ is the qth component of a spherical tensor of rank l that consists of commuting components. In (D.3) $l=0$ for scalar interaction, $l=1$ for spin-rotational interaction and $l=2$ for anisotropic chemical shift, dipolar and quadrupolar interactions. In the course of the discussion of the different interactions it is shown that the relaxation rates are deducible from the spectral density $j(\omega)$ defined by

$$j(\omega) = \frac{1}{2} \int_{-\infty}^{\infty} \langle H_0(0)H_0(t) \rangle \, e^{-i\omega t} \, dt. \qquad (D.4)$$

When $H_q(t)$ depends only on orientational variables we see from (C.35) and (D.4) that

$$j(\omega) = \frac{1}{2(2l+1)} \sum_{n,n'=-l}^{l} H_n'^* H_n' \int_{-\infty}^{\infty} \langle R^+(t) \rangle_{nn'} \, e^{-i\omega t} \, dt. \qquad (D.5)$$

For simplicity we omit the l from the subscripts of H', H'^* since the rank is obvious from the range of the summation.

In order to discuss (D.4) and (D.5) we define the operator $\tau(\omega)$ by

$$\tau(\omega) = \int_0^{\infty} \langle R^+(t) \rangle \, e^{-i\omega t} \, dt. \qquad (D.6)$$

Since $H_0(t)$ is real, it follows that the correlation function in (D.4) is real. Since $H_0(t)$ is a commuting variable and since the random motion is in a steady state,

$$\langle H_0(0)H_0(-t) \rangle = \langle H_0(t)H_0(0) \rangle = \langle H_0(0)H_0(t) \rangle$$

and the correlation function is even. Hence,

$$j(\omega) = \frac{1}{2}\int_0^\infty \langle H_0(0)H_0(t)\rangle\, e^{-i\omega t} + \text{c.c.}, \qquad (D.7)$$

where c.c. denotes complex conjugate. We have from (C.35) and (D.6)

$$\int_0^\infty \langle H_0(0)H_0(t)\rangle\, e^{-i\omega t}\, dt = \frac{1}{2l+1}\sum_{n,n'=-l}^{l} H_n'^{*}H_n'\tau(\omega)_{nn'}. \qquad (D.8)$$

On interchanging the summation indices n, n', affixing the complex conjugate and employing (A.13) we obtain

$$\left(\int_0^\infty \langle H_0(0)H_0(t)\rangle\, e^{-i\omega t}\, dt\right)^{*} = \frac{1}{2l+1}\sum_{n,n'=-l}^{l} H_{n'}'^{*}H_n'\tau(\omega)_{nn'}^{+}. \qquad (D.9)$$

Writing

$$\tau(\omega) + \tau(\omega)^{+} = \sigma(\omega) \qquad (D.10)$$

we deduce from (D.7)–(D.10) that

$$j(\omega) = \frac{1}{2(2l+1)}\sum_{n,n'=-l}^{l} H_n'\sigma(\omega)_{nn'}H_n'^{*}. \qquad (D.11)$$

Hence, in order to obtain the spectral density $j(\omega)$ we must calculate $\langle R^{+}(t)\rangle$, deduce $\sigma(\omega)$ from (D.6) and (D.10), evaluate the matrix elements of $\sigma(\omega)$ in the representation with basis consisting of the spherical harmonics in the molecular coordinate frame of the Euler–Langevin equations, and substitute into (D.11).

Another application of (D.6) is the calculation of correlation times. Let us find the correlation time τ_l for $Y_{lq}(\theta(t), \phi(t))$. According to (2.24)

$$\tau_l = \frac{\int_0^\infty \langle Y_{lq}^{*}(\theta(0), \phi(0))Y_{lq}(\theta(t), \phi(t))\rangle\, dt}{\langle Y_{lq}^{*}(\theta(0), \phi(0))Y_{lq}(\theta(0), \phi(0))\rangle}. \qquad (D.12)$$

Putting $t = 0$, $q' = q$ and $\langle R^{+}(0)\rangle = E$ in (C.31) we have

$$\langle Y_{lq}^{*}(\theta(0), \phi(0))Y_{lq}(\theta(0), \phi(0))\rangle = \frac{1}{2l+1}\sum_{n=-l}^{l}|Y_{ln}(\theta', \phi')|^2 = \frac{1}{4\pi} \qquad (D.13)$$

(Rose, 1957, eq. (4.28)). Then from (C.31), (D.6) and (D.12)

$$\tau_l = \frac{4\pi}{2l+1}\sum_{n,n'=-l}^{l} Y_{ln}(\theta', \phi')\tau(0)_{nn'}Y_{ln'}^{*}(\theta', \phi'). \qquad (D.14)$$

This is called the *orientational correlation time* to distinguish it from the *angular velocity correlation time*, which is deducible from (2.24) when $A(t)$ is a component $\omega_i(t)$ of angular velocity appearing in the Euler–Langevin equations.

In section 3.3 the correlation time τ_l' for the relaxation process described by the interaction Hamiltonian $\hbar G(t)$, where $G(t)$ is given by (3.38) with

$j = l$, is defined as the correlation time for $H_{lq}(t)$. When $H_{lq}(t)$ is a function of orientational variables only, the reasoning employed in establishing (D.14) gives

$$\tau'_l = \frac{\sum_{n,n'=-l}^{l} H'_{ln}\tau(0)_{nn'} H'^*_{ln'}}{\sum_{n=-l}^{l} |H'_{ln}|^2}. \tag{D.15}$$

We see that τ'_l is equal to τ_l, if

$$H_{lq}(\mathbf{a}) = c Y_{lq}(\theta', \phi'), \tag{D.16}$$

where the direction of $\mathbf{a}$ is specified by the angular coordinates θ', ϕ', and c is a constant independent of q.

We shall now derive a relation between $j(\omega)$ and τ'_l in the extreme narrowing approximation where the argument of the spectral density may be replaced by zero. From (D.11)

$$j(0) = \frac{1}{2(2l+1)} \sum_{n,n'=-l}^{l} H'_n \sigma(0)_{nn'} H'^*_{n'}. \tag{D.17}$$

If $\tau(0)_{nn'}$ is a real and symmetric matrix, as it is for all the molecular models that we examine, it will follow from (A.13) and (D.10) that

$$\sigma(0)_{nn'} = 2\tau(0)_{nn'}.$$

Then, from (D.17),

$$j(0) = \frac{1}{2l+1} \sum_{n,n'=-l}^{l} H'_n \tau(0)_{nn'} H'^*_{n'}$$

and (D.15) yields

$$j(0) = \frac{\tau'_l}{2l+1} \sum_{n=-l}^{l} |H'_{ln}|^2. \tag{D.18}$$

Hence, the ratio of $j(0)$ to τ'_l depends only on values of the components of the spherical tensor $H_{ln}(\mathbf{a})$ in the body fixed coordinate system. Now $H_{ln}(\mathbf{a})$ is obtained from the $H_{-q}(t)$ of (D.3) by a rotation of coordinate axes from the laboratory frame to the body frame, so for a prescribed relaxation Hamiltonian $\hbar G$ the ratio of $j(0)$ to τ'_l is independent of the molecular model. Thus, if we calculate the ratio for the spherical model, we know that this is also the ratio for a linear or asymmetric top model.

It is found that the nuclear magnetic relaxation rates T_1^{-1}, T_2^{-1} are expressible as

$$a_0 j(0) + a_1 j(\omega_0) + a_2 j(2\omega_0), \tag{D.19}$$

where a_0, a_1, a_2 are constants. In the extreme narrowing approximation (D.19) reduces to $(a_0 + a_1 + a_2)j(0)$. Then the ratio of a relaxation rate to the correlation time τ'_l is also independent of the molecular model (McConnell, 1986b).

In the rest of the appendix we shall report the values of $\langle R^+(t) \rangle$, $\tau(\omega)_{nn'}$,

$\sigma(\omega)_{nn'}$ for molecules that are spherical, linear, symmetric and asymmetric. The results will be given for each molecular model as a consequence of calculations based on the Euler–Langevin equations (2.57) and therefore including inertial effects. When combining the results with (C.31)–(C.35) we must therefore take for the body fixed axes of the latter equations the principal axes of inertia through the centre of mass of the rotating molecule. Results in rotational diffusion theory will be deduced by applying the limiting process (2.61). We commence with the case of a spherical molecule.

D.1 The spherical molecule

When the molecule is spherical, (2.57) reduce to the simple form

$$I_1 \frac{d\omega(t)}{dt} = -I_1 B\omega(t) + A(t),$$

where I_1 is the moment of inertia about a diameter, $I_1 B$ is the friction coefficient and $A(t)$ is the random driving thermal couple. Then, as pointed out after (2.56), B^{-1} is the friction time τ_F. Moreover (McConnell, 1980a)

$$
\begin{aligned}
\langle R(t) \rangle = & \left[E + \kappa J^2 (1 - e^{-Bt}) + \kappa^2 \{ J^2 [\tfrac{5}{4} - (Bt+1)e^{-Bt} - \tfrac{1}{4}e^{-2Bt}] \right. \\
& + (J^2)^2 [\tfrac{1}{2} - e^{-Bt} + \tfrac{1}{2}e^{-2Bt}] \} + \kappa^3 \{ J^2 [\tfrac{19}{9} - (\tfrac{1}{2}B^2 t^2 + 2Bt + 1)e^{-Bt} \\
& - (\tfrac{3}{4}Bt+1)e^{-2Bt} - \tfrac{1}{9}e^{-3Bt}] + (J^2)^2 [\tfrac{5}{4} - (Bt + \tfrac{9}{4})e^{-Bt} \\
& + (Bt + \tfrac{3}{4})e^{-2Bt} + \tfrac{1}{4}e^{-3Bt}] \\
& + (J^2)^3 [\tfrac{1}{6} - \tfrac{1}{2}e^{-Bt} + \tfrac{1}{2}e^{-2Bt} - \tfrac{1}{6}e^{-3Bt}] \} + \cdots] \\
& \times \exp[-\kappa B \{ 1 + \tfrac{1}{2}\kappa + \tfrac{7}{12}\kappa^2 + (\tfrac{17}{18} - \tfrac{1}{8}J^2)\kappa^3 + \cdots \} J^2 t],
\end{aligned}
$$

(D.20)

where

$$\kappa = \frac{kT}{I_1 B^2} \qquad (D.21)$$

is a small dimensionless constant. The value of $\langle R(t) \rangle$ is clearly a multiple of the identity operator E. For a $(2j+1)$-dimensional representation we replace J^2 by $j(j+1)E$. On putting

$$G' = j(j+1)\kappa \{ 1 + \tfrac{1}{2}\kappa + \tfrac{7}{12}\kappa^2 + [\tfrac{17}{18} - \tfrac{1}{8}j(j+1)]\kappa^3 + \cdots \}$$

$$\frac{\omega}{B} = \omega' \qquad (D.22)$$

and affixing the superscript j to $\tau(\omega)$ and $\sigma(\omega)$ in order to indicate the rank of the representation it is found that

$$\tau(\omega)^j_{mn} = \delta_{mn}\left\{\frac{1}{B(G'+i\omega')} + \frac{j(j+1)\kappa}{B}\left[\frac{1}{G'+i\omega'} - \frac{1}{1+G'+i\omega'}\right]\right.$$

$$+\frac{j(j+1)\kappa^2}{B}\left[\frac{\frac{5}{4}+\frac{1}{2}j(j+1)}{G'+i\omega'} - \frac{1+j(j+1)}{1+G'+i\omega'} - \frac{1}{(1+G'+i\omega')^2}\right.$$

$$\left.\left.+\frac{\frac{1}{2}j(j+1)-\frac{1}{4}}{2+G'+i\omega'}\right]+\cdots\right\} \tag{D.23}$$

$$\sigma(\omega)^j_{mn} = \delta_{mn}\left\{\frac{2G'}{B(G'^2+\omega'^2)}\right.$$

$$\left.+\frac{2j(j+1)\kappa}{B}\frac{G'(1+G')-\omega'^2}{(G'^2+\omega'^2)[(1+G')^2+\omega'^2]}+\cdots\right\}. \tag{D.24}$$

We see from (D.13), (D.14) and (D.23) that

$$\tau_j = \tau(0)^j_{00}. \tag{D.25}$$

To deduce the rotational diffusion limits of these and of later results we perform the limiting operations

$$I_i \to 0, \qquad B_i \to \infty, \qquad I_iB_i \text{ finite} \qquad (i=1,2,3)$$

of (2.61). Then from (D.20)–(D.25)

$$\kappa=0, \qquad BG'=j(j+1)\frac{kT}{I_1B} \tag{D.26}$$

$$\langle R(t)\rangle = \exp\left[-\frac{kT}{I_1B}J^2t\right] \tag{D.27}$$

$$\tau(0)^j_{mn} = \frac{\delta_{mn}}{j(j+1)}\frac{I_1B}{kT} \tag{D.28}$$

$$\sigma(\omega)^j_{mn} = \frac{2\delta_{mn}I_1B}{j(j+1)kT}\frac{1}{1+\left(\frac{I_1B\omega}{j(j+1)kT}\right)^2}. \tag{D.29}$$

When $j=2$ we have

$$\sigma(\omega)^2_{00} = \frac{I_1B}{3kT}\frac{1}{1+\left(\frac{I_1B\omega}{6kT}\right)^2} \tag{D.30}$$

and, from (D.25) and (D.28),

$$\tau_2 = \frac{I_1B}{6kT}. \tag{D.31}$$

D.2 The linear molecule

In the linear rotator model of the molecule we take rotating axes with the origin at the centre, the third axis being along the molecule and the other two axes perpendicular to the molecule and to each other. We denote by I_1 the moment of inertia about axes 1 and 2, by $I_1 B$ the friction coefficient and by A_1, A_2 the components of the random driving couple. It is assumed that the moment of inertia and angular velocity component about the third axis are zero. There are now two Langevin equations

$$I_1 \frac{d\omega_1(t)}{dt} = -I_1 B\omega_1(t) + A_1(t)$$

$$I_1 \frac{d\omega_2(t)}{dt} = -I_1 B\omega_2(t) + A_2(t).$$

Then B^{-1} is the friction time τ_F introduced in subsection 2.4.2, and (McConnell, 1982c)

$$\langle R(t) \rangle = \{E + \kappa(J^2 - J_3^2)(1 - e^{-Bt})$$
$$+ \kappa^2[(2J^2 - 5J_3^2)(\tfrac{5}{4} - Bt\,e^{-Bt} - e^{-Bt} - \tfrac{1}{4}e^{-2Bt})$$
$$+ (J^2 - J_3^2)^2(\tfrac{1}{2} - e^{-Bt} + \tfrac{1}{2}e^{-2Bt})] + \cdots\} \exp(-BGt), \quad (D.32)$$

where

$$G = \kappa\{(J^2 - J_3^2) + \kappa(J^2 - \tfrac{5}{2}J_3^2) + \kappa^2(\tfrac{8}{3}J^2 - \tfrac{29}{3}J_3^2 + 4[J^2 - J_3^2]J_3^2) + \cdots\}. \quad (D.33)$$

In our applications we shall require only the five-dimensional representation, and henceforth we shall omit the superscript 2 to τ and σ. The matrix representative of J^2 is six times the unit matrix and that of J_3^2 is the diagonal matrix with consecutive elements 4, 1, 0, 1, 4. Writing

$$G_0 = 6\kappa(1 + \kappa + \tfrac{8}{3}\kappa^2 + \cdots) \quad (D.34)$$
$$G_1 = 5\kappa(1 + \tfrac{7}{10}\kappa + \tfrac{79}{15}\kappa^2 + \cdots) \quad (D.35)$$
$$G_2 = 2\kappa(1 - 2\kappa + \tfrac{14}{3}\kappa^2 + \cdots) \quad (D.36)$$

one finds that the non-vanishing elements of the matrix representatives of $\tau(\omega)$ and $\sigma(\omega)$ are the following (McConnell, 1983):

$$\tau(\omega)_{00} = \frac{1}{B}\left\{\frac{1 + 6\kappa + 33\kappa^2 + \cdots}{G_0 + i\omega'} - \frac{6\kappa + 48\kappa^2 + \cdots}{1 + G_0 + i\omega'} \right.$$
$$\left. - \frac{12\kappa^2 + \cdots}{(1 + G_0 + i\omega')^2} + \frac{15\kappa^2 + \cdots}{2 + G_0 + i\omega'} + \cdots\right\} \quad (D.37)$$

$$\tau(\omega)_{11} = \tau(\omega)_{-1,-1} = \frac{1}{B} \left\{ \frac{1 + 5\kappa + \frac{85}{4}\kappa^2 + \cdots}{G_1 + i\omega'} - \frac{5\kappa + 32\kappa^2 + \cdots}{1 + G_1 + i\omega'} \right.$$

$$\left. - \frac{7\kappa^2 + \cdots}{(1 + G_1 + i\omega')^2} + \frac{\frac{43}{4}\kappa^2 + \cdots}{2 + G_1 + i\omega'} + \cdots \right\} \quad \text{(D.38)}$$

$$\tau(\omega)_{22} = \tau(\omega)_{-2,-2} = \frac{1}{B} \left\{ \frac{1 + 2\kappa - 8\kappa^2 + \cdots}{G_2 + i\omega'} + \frac{-2\kappa + 4\kappa^2 + \cdots}{1 + G_2 + i\omega'} \right.$$

$$\left. + \frac{8\kappa^2 + \cdots}{(1 + G_2 + i\omega')^2} + \frac{4\kappa^2 + \cdots}{2 + G_2 + i\omega'} + \cdots \right\}$$
$$\text{(D.39)}$$

$$\sigma(\omega)_{00} = \frac{1}{B} \left\{ \frac{2(1 + 6\kappa + 33\kappa^2 + \cdots)G_0}{G_0^2 + \omega'^2} - \frac{2(6\kappa + 48\kappa^2 + \cdots)(1 + G_0)}{(1 + G_0)^2 + \omega'^2} \right.$$

$$\left. - \frac{24\kappa^2[(1 + G_0)^2 - \omega'^2]}{[(1 + G_0)^2 + \omega'^2]^2} + \frac{30\kappa^2(2 + G_0)}{(2 + G_0)^2 + \omega'^2} + \cdots \right\} \quad \text{(D.40)}$$

$$\sigma(\omega)_{11} = \sigma(\omega)_{-1,-1} = \frac{1}{B} \left\{ \frac{2(1 + 5\kappa + \frac{85}{4}\kappa^2 + \cdots)G_1}{G_1^2 + \omega'^2} - \frac{2(5\kappa + 32\kappa^2 + \cdots)(1 + G_1)}{(1 + G_1)^2 + \omega'^2} \right.$$

$$\left. - \frac{14\kappa^2[(1 + G_1)^2 - \omega'^2]}{[(1 + G_1)^2 + \omega'^2]^2} + \frac{\frac{43}{2}\kappa^2(2 + G_1)}{(2 + G_1)^2 + \omega'^2} + \cdots \right\}$$
$$\text{(D.41)}$$

$$\sigma(\omega)_{22} = \sigma(\omega)_{-2,-2} = \frac{1}{B} \left\{ \frac{2(1 + 2\kappa - 8\kappa^2 + \cdots)G_2}{G_2^2 + \omega'^2} + \frac{2(-2\kappa + 4\kappa^2 + \cdots)}{(1 + G_2)^2 + \omega'^2} \right.$$

$$\left. + \frac{16\kappa^2[(1 + G_2)^2 - \omega'^2]}{[(1 + G_2)^2 + \omega'^2]^2} + \frac{8\kappa^2(2 + G_2)}{(2 + G_2)^2 + \omega'^2} + \cdots \right\}.$$
$$\text{(D.42)}$$

In the rotational diffusion limit (2.61), (D.32) and (D.33) yield

$$\langle R(t) \rangle = \exp\left[-\frac{kT}{I_1 B}(J^2 - J_3^2)t \right]. \quad \text{(D.43)}$$

Then from (D.34)–(D.36)

$$G_0 = 6\kappa, \qquad G_1 = 5\kappa, \qquad G_2 = 2\kappa. \quad \text{(D.44)}$$

We deduce from (D.37)–(D.42) and (D.44) that

$$\tau(0)_{00} = \frac{I_1 B}{6kT}, \qquad \tau(0)_{11} = \tau(0)_{-1,-1} = \frac{I_1 B}{5kT}, \qquad \tau(0) = \tau(0)_{-2,-2} = \frac{I_1 B}{2kT}$$
$$\text{(D.45)}$$

$$\sigma(\omega)_{00} = \frac{I_1 B}{3kT\left[1 + \left(\frac{I_1 B\omega}{6kT}\right)^2\right]} \tag{D.46}$$

$$\sigma(\omega)_{11} = \sigma(\omega)_{-1,-1} = \frac{2I_1 B}{5kT\left[1 + \left(\frac{I_1 B\omega}{5kT}\right)^2\right]} \tag{D.47}$$

$$\sigma(\omega)_{22} = \sigma(\omega)_{-2,-2} = \frac{I_1 B}{kT\left[1 + \left(\frac{I_1 B\omega}{2kT}\right)^2\right]} \tag{D.48}$$

D.3 The asymmetric molecule

The rotational Brownian motion of an asymmetric rotator has been investigated by Ford *et al.* (1979). The central result expressed in our notation is (McConnell, 1980*b*, Chap. 11)

$$\langle R(t)\rangle = \left\{E + \sum_{i=1}^{3} \frac{kT}{I_i B_i^2}(1 - e^{-B_i t})J_i^2 + \cdots\right\}$$

$$\times \exp\left\{\left(-\sum_{i=1}^{3} D_i J_i^2 + \frac{i}{3}P\frac{(kT)^2}{I_1 I_2 I_3}\sum_{i=1}^{3}\frac{B_j - B_k}{B_j^2 B_k^2} + \cdots\right)t\right\}, \tag{D.49}$$

where i, j, k is a cyclic permutation of 1, 2, 3, D_i is $kT/(I_i B_i)$ plus a small correction of relative order κ defined in (D.21), and

$$P = J_1 J_2 J_3 + J_1 J_3 J_2 + J_2 J_3 J_1 + J_2 J_1 J_3 + J_3 J_1 J_2 + J_3 J_2 J_1.$$

It is deduced from (D.6) that for $j = 2$

$$\tau(\omega)_{mn} = \begin{bmatrix} A & 0 & D & 0 & F \\ 0 & B & 0 & E & 0 \\ D & 0 & C & 0 & D \\ 0 & E & 0 & B & 0 \\ F & 0 & D & 0 & A \end{bmatrix}, \tag{D.50}$$

where

$$A = \frac{(ac - d^2)(1 + \tilde{a}) - ad\tilde{d}}{a(ac - 2d^2)}, \qquad B = \frac{b + b\tilde{b} - e\tilde{e}}{b^2 - e^2}$$

$$C = \frac{a + a\tilde{c} - 2d\tilde{d}}{ac - 2d^2}, \qquad D = \frac{-2d + (a+c)\tilde{d} - d(\tilde{a}+\tilde{c})}{2(ac - 2d^2)} \tag{D.51}$$

$$E = \frac{-e + b\tilde{e} - e\tilde{b}}{b^2 - e^2}, \qquad F = \frac{d^2 + d(d\tilde{a} - a\tilde{d})}{a(ac - 2d^2)}$$

$$a = D_1 + D_2 + 4D_3 + i\omega, \qquad b = \tfrac{5}{2}(D_1 + D_2) + D_3 + i\omega$$
$$c = 3(D_1 + D_2) + i\omega, \qquad d = (\tfrac{2}{3})^{1/2}e = (\tfrac{3}{2})^{1/2}(D_1 - D_2)$$

$$\tilde{a} = \frac{D_1}{B_1} + \frac{D_2}{B_2} + \frac{4D_3}{B_3}, \qquad \tilde{b} = \frac{5}{2}\left(\frac{D_1}{B_1} + \frac{D_2}{B_2}\right) + \frac{D_3}{B_3}$$

$$\tilde{c} = 3\left(\frac{D_1}{B_1} + \frac{D_2}{B_2}\right), \qquad \tilde{d} = (\tfrac{2}{3})^{1/2}\tilde{e} = (\tfrac{3}{2})^{1/2}\left(\frac{D_1}{B_1} - \frac{D_2}{B_2}\right). \tag{D.52}$$

Since the matrix in (D.50) is symmetric, we deduce from (D.10) that

$$\sigma(\omega)_{mn} = \begin{bmatrix} A+A^* & 0 & D+D^* & 0 & F+F^* \\ 0 & B+B^* & 0 & E+E^* & 0 \\ D+D^* & 0 & C+C^* & 0 & D+D^* \\ 0 & E+E^* & 0 & B+B^* & 0 \\ F+F^* & 0 & D+D^* & 0 & A+A^* \end{bmatrix} \cdot \tag{D.53}$$

In the rotational diffusion limit (D.49) reduces to

$$\langle R(t) \rangle = \exp(G_D t), \tag{D.54}$$

where

$$G_D = -\left(\frac{kT}{I_1 B_1} J_1^2 + \frac{kT}{I_2 B_2} J_2^2 + \frac{kT}{I_3 B_3} J_3^2\right), \tag{D.55}$$

the small correction to $kT/(I_i B_i)$ in the value of D_i now vanishing. From (D.6) and (D.54)

$$\tau(\omega) = \int_0^\infty \exp[(G_D - i\omega E)t] \, dt.$$

On integrating it is found that (McConnell, 1982a)

$$\tau(\omega) = (-G_D + i\omega E)^{-1}.$$

The values of $\tilde{a}, \tilde{b}, \tilde{c}, \tilde{d}, \tilde{e}$ are, from (D.52), of order κ and therefore negligible. Hence, it is deduced from (D.50) and (D.51) that the matrix representative $\tau(\omega)_{mn}$ is given by

$$\tau(\omega)_{mn} = \begin{bmatrix} A' & 0 & D' & 0 & F' \\ 0 & B' & 0 & E' & 0 \\ D' & 0 & C' & 0 & D' \\ 0 & E' & 0 & B' & 0 \\ F' & 0 & D' & 0 & A' \end{bmatrix}, \tag{D.56}$$

where, by (D.51),

$$A' = \frac{ac - d^2}{a(ac - 2d^2)}, \qquad B' = \frac{b}{b^2 - e^2}, \qquad C' = \frac{a}{ac - 2d^2}$$

$$D' = -\frac{d}{ac - 2d^2}, \qquad E' = -\frac{e}{b^2 - e^2}, \qquad F' = \frac{d^2}{a(ac - 2d^2)} \qquad \text{(D.57)}$$

and a, b, c, d, e are defined in (D.52). The matrix representative of $\sigma(\omega)$ is obtained by making in (D.56) the substitutions

$$\begin{aligned} A' &\mapsto A' + A'^*, & B' &\mapsto B' + B'^*, & C' &\mapsto C' + C'^* \\ D' &\mapsto D' + D'^*, & E' &\mapsto E' + E'^*, & F' &\mapsto F' + F'^*. \end{aligned} \qquad \text{(D.58)}$$

D.4 The symmetric molecule

When the molecule is a symmetric rotator, we put $I_2 = I_1, B_2 = B_1, D_2 = D_1$ in (2.57) and (D.49). It follows from (D.51) and (D.52) that

$$d = e = \tilde{d} = \tilde{e} = 0, \qquad D = E = F = 0,$$

$$a = 2D_1 + 4D_3 + i\omega, \qquad b = 5D_1 + D_3 + i\omega, \qquad c = 6D_1 + i\omega$$

$$\tilde{a} = \frac{2D_1}{B_1} + \frac{4D_3}{B_3}, \qquad \tilde{b} = \frac{5D_1}{B_1} + \frac{D_3}{B_3}, \qquad \tilde{c} = \frac{6D_1}{B_1} \qquad \text{(D.59)}$$

$$A = \frac{1 + 2D_1/B_1 + 4D_3/B_3}{2D_1 + 4D_3 + i\omega}, \quad B = \frac{1 + 5D_1/B_1 + D_3/B_3}{5D_1 + D_3 + i\omega}, \quad C = \frac{1 + 6D_1/B_1}{6D_1 + i\omega}.$$

We see from (D.50) and (D.53) that the matrix representatives of $\tau(\omega)$ and $\sigma(\omega)$ are diagonal.

In the rotational diffusion limit we deduce from (D.57) that

$$A' = \frac{1}{a}, \qquad B' = \frac{1}{b}, \qquad C' = \frac{1}{c}, \qquad D' = E' = F' = 0,$$

where a, b, c are given by (D.59) and now

$$D_1 = \frac{kT}{I_1 B_1}, \qquad D_3 = \frac{kT}{I_3 B_3}. \qquad \text{(D.60)}$$

From (D.56) it follows that $\tau(\omega)$ is represented by a five-dimensional diagonal matrix with elements

$$\tau(\omega)_{00} = \left(\frac{6kT}{I_1 B_1} + i\omega \right)^{-1} \qquad \text{(D.61)}$$

$$\tau(\omega)_{11} = \tau(\omega)_{-1,-1} = \left(\frac{5kT}{I_1 B_1} + \frac{kT}{I_3 B_3} + i\omega\right)^{-1} \qquad \text{(D.62)}$$

$$\tau(\omega)_{22} = \tau(\omega)_{-2,-2} = \left(\frac{2kT}{I_1 B_1} + \frac{4kT}{I_3 B_3} + i\omega\right)^{-1} \qquad \text{(D.63)}$$

Similarly $\sigma(\omega)$ is represented by a five-dimensional diagonal matrix with elements

$$\sigma(\omega)_{00} = \frac{I_1 B_1}{3kT[1 + (I_1 B_1 \omega/6kT)^2]} \qquad \text{(D.64)}$$

$$\sigma(\omega)_{11} = \sigma(\omega)_{-1,-1} = \frac{2[5kT/(I_1 B_1) + kT/(I_3 B_3)]}{[5kT/(I_1 B_1) + kT/(I_3 B_3)]^2 + \omega^2} \qquad \text{(D.65)}$$

$$\sigma(\omega)_{22} = \sigma(\omega)_{-2,-2} = \frac{4[kT/(I_1 B_1) + 2kT/(I_3 B_3)]}{4[kT/(I_1 B_1) + 2kT/(I_3 B_3)]^2 + \omega^2}. \qquad \text{(D.66)}$$

If we compare (D.30), (D.46) and (D.64), we see that in the Debye approximation $\sigma(\omega)_{00}$ has the same expression for the sphere, the linear rotator and the symmetric rotator, the quantity B_1 in the last case clearly corresponding to B in the other two cases. Thus, when we have a problem in which only $\sigma(\omega)_{00}$ makes a contribution to $j(\omega)$ in (D.11), the spectral density will have the same value in the three molecular models. This will occur, for example, if $H'_n \equiv Y_{ln}(\theta', \phi')$ and θ' is zero; indeed we see from (B.1) that the spherical harmonics then present in (D.11) will vanish except when $n = n' = 0$ (McConnell, 1986b).

References

Abragam, A. (1961). *The Principles of Nuclear Magnetism.* Oxford: The Clarendon Press.

Albrand, J. P., Taieb, M. C., Fries, P. & Belorizky, E. (1981). Effects of eccentricity on nuclear magnetic relaxation by intermolecular dipole–dipole interactions: ^{13}C relaxation of neopentane. *The Journal of Chemical Physics,* 75, 2141–6.

Albrand, J. P., Taieb, M. C., Fries, P. H. & Belorizky, E. (1983). NMR study of spectral densities over a large frequency range for intermolecular relaxation in liquids: Pair correlation effects. *The Journal of Chemical Physics,* 78, 5809–15.

Andrew, E. R. (1969). *Nuclear Magnetic Resonance.* Cambridge University Press.

Atherton, N. M. (1973). *Electron Spin Resonance.* Chichester: Ellis Horwood.

Ayant, Y., Belorizky, E., Alizon, J. & Gallice, J. (1975). Calcul des densités spectrales résultant d'un mouvement aléatoire de translation en relaxation par interaction dipolaire magnétique dans les liquides. *Le Journal de Physique,* 36, 991–1004.

Ayant, Y., Belorizky, E., Fries, P. & Rosset, J. (1977). Effet des interactions dipolaires magnétiques intermoléculaires sur la rélaxation nucléaire de molécules polyatomiques dans les liquides. *Le Journal de Physique,* 38, 325–37.

Batchelor, G. K. (1967). *An Introduction to Fluid Dynamics.* Cambridge University Press.

Bender, H. J. & Zeidler, M. D. (1971). Translational and anisotropic rotational diffusion in liquid chloroform as studied by NMR relaxation. *Berichte der Bunsengessellschaft für Physikalische Chemie,* 75, 236–42.

Blicharska, B., Frech, T. & Hertz, H. G. (1984). Rotational, internal rotational and translational motion of dimethyl sulfoxide in its liquid mixture with water. *Zeitschrift für Physikalische Chemie.* Neue Folge, 141, 139–57.

Bloch, F. (1946). Nuclear induction. *Physical Review,* 70, 460–74.

Bloembergen, N., Purcell, E. M. & Pound, R. V. (1948). Relaxation effects in nuclear magnetic resonance absorption. *Physical Review,* 73, 679–712.

Bonera, G. & Rigamonti, A. (1965a). Intra- and intermolecular contributions to the proton spin-lattice relaxation in liquids. *The Journal of Chemical Physics,* 42, 171–4.

Bonera, G. & Rigamonti, A. (1965b). Electric-field gradients in liquids by deuteron quadrupole relaxation. *The Journal of Chemical Physics,* 42, 175–80.

Born, M. (1957). *Atomic Physics,* 6th edn. London: Blackie.

Chandrasekhar, S. (1943). Stochastic problems in physics and astronomy. *Reviews of Modern Physics,* 15, 1–89.

Condon, E. U. & Shortley, G. H. (1935). *The Theory of Atomic Spectra.* Cambridge University Press.

Curie, P. (1895). Propriétes magnétiques des corps à diverses températures. *Annales de Chimie et de Physique* (7) 5, 289–405

Debye, P. (1929). *Polar Molecules.* New York: Chemical Catalog Company.

De Graaf, L. A. (1969). Study of molecular motions in cyclohexane and cyclopentane in cold-neutron scattering. *Physica* 40, 497–516.

Eckart, C. (1930). The application of group theory to the quantum dynamics of monatomic systems. *Reviews of Modern Physics*, 2, 305–80.

Edmonds, A. R. (1968). *Angular Momentum in Quantum Mechanics*, 2nd edn. Princeton University Press.

Farrar, T. C. & Becker, E. D. (1971). *Pulse and Fourier Transform NMR*. New York: Academic Press.

Farrar, T. C., Maryott, A. A. & Malmberg, M. S. (1972). Nuclear magnetic resonance. *Annual Review of Physical Chemistry*, 23, 193–216.

Ford, G. W., Lewis, J. T. & McConnell, J. (1979). Rotational Brownian motion of an asymmetric top. *Physical Review*, A 19, 907–19.

Frech, T. & Hertz, H. G. (1984). On the rotational correlation function $g_2(t)$ of the water molecule in a water–dimethylsulfoxide mixture. *Zeitschrift für Physikalische Chemie, Neue Folge*, 142, 43–65.

Frech, T. & Hertz, H. G. (1985). Rotational, internal rotational and translational motion of acetic acid and dimethyl sulfoxide in their liquid mixture. *Journal of Molecular Liquids*, 30, 237–82.

Gordon, R.G . (1966). On the rotational diffusion of molecules. *The Journal of Chemical Physics*, 44, 1830–6.

Goulon, J., Rivail, J. L., Fleming, J. W., Chamberlain, J. & Chantry, G. W. (1973). Dielectric relaxation and far infrared dispersion in pure liquid chloroform. *Chemical Physics Letters*, 18, 211–16.

Gutowsky, H. S., Lawrenson, I. J. & Shimomura, K. (1961). Nuclear magnetic spin-lattice relaxation by spin-rotational interactions. *Physical Review Letters*, 6, 349–51.

Harmon, J. F. & Muller, B. H. (1969). Nuclear spin relaxation by translational diffusion in liquid ethane. *Physical Review*, 182, 400–10.

Hasted, J. B. (1973). *Aqueous Dielectrics*. London: Chapman Hall.

Heitler, W. (1954). *The Quantum Theory of Radiation*, 3rd ed. Oxford: The Clarendon Press.

Hertz, H. G. (1967). Microdynamic behaviour of liquids as studied by NMR relaxation times. *Progress in Nuclear Magnetic Resonance Spectroscopy*, 3, 159–230.

Hertz, H. G. (1983). The problem of intramolecular rotation in liquids and nuclear magnetic relaxation. *Progress in Nuclear Magnetic Resonance Spectroscopy*, 16, 115–62.

Herzberg, G. (1945). *Molecular Spectra and Molecular Structure II. Infrared and Raman Spectra of Polyatomic Molecules*. New York: Van Nostrand Reinhold Co.

Herzberg, G. (1966). *Molecular Spectra and Molecular Structure III. Electronic Spectra and Electronic Structure of Polyatomic Molecules*. New York: Van Nostrand Reinhold Co.

Herzfeld, K. F. (1964). The absorption of dipole liquids in the visible. *Journal of the American Chemical Society*, 86, 3468–9.

Hubbard, P. S. (1963a). Nuclear magnetic relaxation by intermolecular dipole–dipole interactions. *Physical Review*, 131, 275–82.

Hubbard, P. S. (1963b). Theory of nuclear magnetic relaxation by spin-rotational interactions in liquids. *Physical Review*, 131, 1155–65.

Hubbard, P. S. (1966). Theory of electron-nucleus Overhauser effects in liquids containing free radicals. *Proceedings of the Royal Society of London*, A 291, 537–55.

Hubbard, P. S. (1969). Some properties of correlation functions of irreducible tensor operators. *Physical Review*, 180, 319–26.

Hubbard, P. S. (1974). Nuclear magnetic relaxation in spherical-top molecules undergoing rotational Brownian motion. *Physical Review* A 9, 481–94.

Hubbard, P. S. (1981). Erratum: Nuclear magnetic relaxation in spherical-top molecules undergoing Brownian motion. *Physical Review* A 24, 645.

Huntress, W. T. (1968). Effects of anisotropic molecular rotational diffusion on nuclear magnetic relaxation in liquids. *The Journal of Chemical Physics*, 48, 3524–33.

Huntress, W. T. (1969). A nuclear magnetic resonance study of anisotropic molecular rotation in liquid chloroform and in chloroform–benzene solution. *The Journal of Physical Chemistry*, 75, 103–11.

Jeans, J. H. (1933). *The Mathematical Theory of Electricity and Magnetism*, 5th edn. Cambridge University Press.

Jeffreys, H. (1967). *Theory of Probability*, 3rd edn. Oxford: The Clarendon Press.

Jones, G. P. (1966). Spin-lattice relaxation in the rotating frame: weak-collision case. *Physical Review*, 148, 332–5.

Lamb, H. (1932). *Hydrodynamics*, 6th edn. Cambridge University Press.

Landau, L. D. & Lifschitz, E. M. (1958). *Statistical Physics*. London: Pergamon Press.

Langevin, P. (1905). Magnétisme et théorie des electrons. In *Oeuvres Scientifiques de Paul Langevin*, pp. 331–68. Paris: CNRS, 1950.

Larsson, K. E. (1968). Rotational and translational diffusion in complex liquids. *Physical Review*, 167, 171–82.

Leroy, Y., Constant, E. & Desplanques, P. (1967). Sur l'insuffisance des théories de la relaxation dans l'interprétation des spectres hertziens et ultrahertziens des liquides polaires. *Journal de Chimie Physique et de Physico-Chimie Biologique*, 64, 1499–508.

Littlewood, D. E. (1950). *A University Algebra*. London: Heinemann.

Magnus, W. & Oberhettinger, F. (1949). *Formulas and Theorems for the Special Functions of Mathematical Physics*. New York: Chelsea Publishing Company.

McConnell, H. M. & Holm, C. H. (1956). Anisotropic chemical shielding and nuclear magnetic relaxation in liquids. *The Journal of Chemical Physics*, 25, 1289.

McConnell, J. (1960). *Quantum Particle Dynamics*, 2nd edn. Amsterdam: North-Holland Publishing Company.

McConnell, J. (1980a). Spectral densities of spherical harmonics for rotational Brownian motion. *Physica*, 102A, 539–46.

McConnell, J. (1980b). *Rotational Brownian Motion and Dielectric Theory*. London: Academic Press.

McConnell, J. (1982a). Stochastic differential equation study of nuclear magnetic relaxation by spin-rotational interaction. *Physica*, 111A, 85–113.

McConnell, J. (1982b). Nuclear magnetic spin-rotational relaxation times for symmetric molecules. *Physica*, 112A, 479–87.

McConnell, J. (1982c). Nuclear magnetic spin-rotational relaxation times for linear molecules. *Physica*, 112A, 488–504.

McConnell, J. (1983). Nuclear magnetic relaxation by quadrupole interactions in non-spherical molecules. *Physica*, 117A, 251–64.

McConnell, J. (1984a). Theory of nuclear magnetic relaxation by anisotropic chemical shift. *Physica*, 127A, 152–72.

McConnell, J. (1984b). Debye limit of the stochastic rotation operator. *Physica*, 128A, 611–30.

McConnell, J. (1986a). Theory of nuclear magnetic relaxation by dipolar interaction. *Physica*, 135A, 38–62.

McConnell, J. (1986b). Correlation and nuclear magnetic relaxation times. *Physica*, 138A, 367–81.

McLachlan, N. W. (1934). *Bessel Functions for Engineers*. Oxford University Press.

Peirce, B. O. (1929). *A Short Table of Integrals*. Boston: Ginn and Company.

Pople, J. A., Schneider, W. G. & Bernstein, H. J. (1959). *High-resolution Nuclear Magnetic Resonance*. New York: McGraw-Hill Book Company.

Racah, G. (1962). Theory of complex spectra II. *Physical Review*, 62, 438–62.

Ramsey, N. F. (1950). Magnetic shielding of nuclei in molecules. *Physical Review*, 78, 699–703.

Ramsey, N. F. (1953). *Nuclear Moments*. New York: John Wiley.

Ramsey, N. F. & Purcell, E. M. (1952). Interactions between nuclear spins in molecules. *Physical Review*, 85, 143–4.

Redfield, A. G. (1955). Nuclear magnetic resonance and rotary saturation in solids. *Physical Review*, 98, 1787–809.

Redfield, A. G. (1957). On the theory of relaxation processes. *IBM Journal of Research and Development*, 1, 19–31.

Richter, H. & Zeidler, M. D. (1985). NMR relaxation in 18-crown-6 ether. *Molecular Physics*, 55, 49–59.

Rose, M. E. (1957). *Elementary Theory of Angular Momentum*. New York: John Wiley.

Royal Society (1975). *Quantities, Units, and Symbols*, 2nd edn. London.

Rugheimer, J. H. & Hubbard, P. S. (1963). Nuclear magnetic relaxation and diffusion in liquid CH_4, CF_4, and mixtures of CH_4 and CF_4 with argon. *The Journal of Chemical Physics*, 39, 552–64.

Schiff, L. I. (1968). *Quantum Mechanics*, 3rd edn. New York: McGraw-Hill.

Shimizu, H. (1964). Effect of molecular shape on nuclear magnetic relaxation. II. *The Journal of Chemical Physics*, 40, 754–61.

von Smoluchowski, M. (1915). Über Brownsche Molekularbewegung unter Einwirkung äusserer Kräfte und deren Zusammenhang mit der verallgemeinerten Diffusiongleichung. *Annalen der Physik* (4) 48, 1103–12.

Solomon, I. (1955). Relaxation processes in a system of two spins. *Physical Review*, 99, 559–65.

Solomon, I. & Bloembergen, N. (1956). Nuclear magnetic interactions in the HF molecule. *The Journal of Chemical Physics*, 25, 261–6.

Spiess, H. W., Schweitzer, D., Haeberlen, U. & Hausser, K. H. (1971). Spin-rotation interaction and anisotropic chemical shift in $^{13}CS_2$. *Journal of Magnetic Resonance*, 5, 101–8.

Synge J. L. & Griffith, B. A. (1959). *Principles of Mechanics*, 3rd edn. New York: McGraw-Hill.

Torrey, H. C. (1953). Nuclear spin-relaxation by translation diffusion. *Physical Review*, 92, 962–9.

Townes, C. H. & Schawlow, A. L. (1975). *Microwave Spectroscopy*. New York: Dover Publications.

Van Vleck, J. H. (1951). The coupling of angular momentum vectors in molecules. *Reviews of Modern Physics*, 23, 213–27.

Versmold, H. (1970). Time correlation functions for internal and anisotropic rotational motion of molecules. *Zeitschrift für Naturforschung*, A 25, 367–72.

Wallach, D. (1967). Effect of internal rotation on angular corelation functions. *The Journal of Chemical Physics*, 47, 5258–68.

Wangsness, R. K. & Bloch, F. (1953). The dynamical theory of nuclear induction. *Physical Review*, 89, 728–39.

Wigner, E. P. (1959). *Group Theory and its Applications to the Quantum Mechanics of Atomic Spectra*. New York: Academic Press.

Woessner, D. E. (1962a). Spin relaxation processes in a two-proton system undergoing anisotropic reorientation. *The Journal of Chemical Physics*, 36, 1–4.

Woessner, D.E . (1962b). Nuclear spin relaxaton in ellipsoids undergoing rotational Brownian motion. *The Journal of Chemical Physics*, 37, 647–54.

Zeidler, M. D. (1965). Umorientierungszeiten, Sprungzeiten und Quadrupolkopplungskonstanten in einigen organischen Flüssigkeiten aus kernmagnetischen Relaxationszeitmessungen. *Berichte der Bunsengesellschaft für Physikalische Chemie*, 69, 659–69.

Zeidler, M. D. (1971). A comparative study of quasielastic neutron scattering and NMR relaxation in liquid acetonitrile. *Berichte der Bunsengesellschaft für Physikalische Chemie*, 75, 769–76.

Zeidler, M. D. (1974). Comparative study of molecular motion in liquids by NMR and neutron spectroscopy. In *Molecular Motions in Liquids*, ed. J. Lascombe, pp. 421–38. Dordrecht: D. Reidel.

Zeidler, M. D. (1975). On the theory of nuclear magnetic relaxation by intermolecular dipole–dipole interaction. *Molecular Physics*, 30, 1441–51.

Author index

Subject index

adjoint operator, 159
angular velocity correlation function, 29
angular velocity correlation time, 29, 177
anisotropic chemical shift, 101
asymmetric molecules, 83, 84, 109, 110,
    124, 139–41, 183–5
asymmetry parameter, 104, 109, 122, 124,
    150, 157
autocorrelation function, 22

basis of a representation, 161
benzene C_6H_6, 156
Bloch equations, 7–9, 46
Bohr magneton, 1
Boltzmann distribution, 3, 5
Born's probability density, 32

canonical transformation, *see* unitary
    transformation
carbon disulphide CS_2, 155
carbon tetrachloride CCl_4, 155
carbon tetrafluoride CF_4, 150
centred random variable, 18
chemical exchange, 93–7
chemical exchange time, 94, 148, 149
chemical shift, 101
chloroform $CHCl_3$, 89, 151, 152
circular plate molecules, 85
Clebsch–Gordan coefficients, 116, 170
Cole–Davidson distribution, 158
complete set, 160
complex magnetic susceptibility, 10–12
conditional probability density, 19–21
correlation function, 21
    see also angular velocity correlation
    function, autocorrelation function,
    cross-correlation function, normalized
    autocorrelation function, orientational
    correlation function
correlation time, 22, 39, 148, 149
    see also angular velocity correlation
    time, orientational correlation time
cross-correlation function, 22
crown ether, 151
Curie's law, 5

Debye approximation, 30
    see also asymmetric molecules, circular
    plate molecules, linear molecules,
    planar molecules, spherical molecules,
    symmetric rotator molecules
Debye limit, *see* Debye approximation
Debye theory, 14, 30, 89
density matrix, 33, 57
density operator, 33, 35, 57
diamagnetic screening, 101
diffusion coefficient, *see* rotational diffusion
    coefficient, self-diffusion coefficient
diffusion equation, 27, 81
diffusion theory, 30
diffusive motion, 27
dilution technique, 144
dimethylsulphoxide $(CH_3)_2SO$, 158
Dirac delta function, 23
ditertiobutyl nitroxide $[(CH_3)_3C]_2NO$, 152

effective field, 9
eigenfunction, 159
eigenvalue, 159
electric field gradient tensor, 122, 156, 157
electric quadrupole moment, 115, 118, 156
electron paramagnetic susceptibility, 5
ensemble average, 18, 21, 33
ergodic theorem, 18
Euler angles, 21, 165, 166, 172
Euler–Langevin equations, 29, 175
expectation value, 32
extended diffusion, 30
extreme narrowing approximation, 60,
    88–91, 96, 97, 107, 130
    see also asymmetric molecules, circular
    plate molecules, linear molecules,
    planar molecules, spherical molecules,
    symmetric rotator molecules

field gradient tensor, *see* electric field
    gradient tensor
friction time, 29, 148, 149